Research Notes in Mathematics

Main Editors
H. Brezis, Université de Paris
R. G. Douglas, State University of New York at Stony Brook
A. Jeffrey, University of Newcastle-upon-Tyne *(Founding Editor)*

Editorial Board
R. Aris, University of Minnesota
A. Bensoussan, INRIA, France
W. Bürger, Universität Karlsruhe
J. Douglas Jr, University of Chicago
R. J. Elliott, University of Hull
G. Fichera, Università di Roma
R. P. Gilbert, University of Delaware
R. Glowinski, Université de Paris
K. P. Hadeler, Universität Tübingen
K. Kirchgässner, Universität Stuttgart
B. Lawson, State University of New York at Stony Brook
W. F. Lucas, Cornell University
R. E. Meyer, University of Wisconsin-Madison
J. Nitsche, Universität Freiburg
L. E. Payne, Cornell University
G. F. Roach, University of Strathclyde
J. H. Seinfeld, California Institute of Technology
I. N. Stewart, University of Warwick
S. J. Taylor, University of Virginia

Submission of proposals for consideration

Suggestions for publication, in the form of outlines and representative samples, are invited by the Editorial Board for assessment. Intending authors should approach one of the main editors or another member of the Editorial Board, citing the relevant AMS subject classifications. Alternatively, outlines may be sent directly to one of the publisher's offices. Refereeing is by members of the board and other mathematical authorities in the topic concerned, throughout the world.

Preparation of accepted manuscripts

On acceptance of a proposal, the publisher will supply full instructions for the preparation of manuscripts in a form suitable for direct photo-lithographic reproduction. Specially printed grid sheets are provided and a contribution is offered by the publisher towards the cost of typing. Word processor output, subject to the publisher's approval, is also acceptable.

Illustrations should be prepared by the authors, ready for direct reproduction without further improvement. The use of hand-drawn symbols should be avoided wherever possible, in order to maintain maximum clarity of the text.

The publisher will be pleased to give any guidance necessary during the preparation of a typescript, and will be happy to answer any queries.

Important note

In order to avoid later retyping, intending authors are strongly urged not to begin final preparation of a typescript before receiving the publisher's guidelines and special paper. In this way it is hoped to preserve the uniform appearance of the series.

Advanced Publishing Program
Pitman Publishing Inc
1020 Plain Street
Marshfield, MA 02050, USA
(tel (617) 837 1331)

Advanced Publishing Program
Pitman Publishing Limited
128 Long Acre
London WC2E 9AN, UK
(tel 01-379 7383)

Hamilton's principle in continuum mechanics

A Bedford

University of Texas at Austin

Hamilton's principle in continuum mechanics

Pitman Advanced Publishing Program
BOSTON · LONDON · MELBOURNE

PITMAN PUBLISHING LIMITED
128 Long Acre, London WC2E 9AN

A Longman Group Company

© A Bedford 1985

First published 1985

AMS Subject Classifications: (main) 49H05, 70H25, 73A05, 76T05
(subsidiary) 73B99, 73C02, 73C50, 73G10, 73S99

ISSN 0743-0337

Library of Congress Cataloging in Publication Data

Bedford, A.
   Hamilton's principle in continuum mechanics.

   Bibliography: p.
   1. Continuum mechanics.   2. Hamilton, William Rowan,
Sir, 1805–1865.   I. Title.
QA808.2.B385   1985   530.1'2   85-24448
ISBN 0-273-08730-4

British Library Cataloguing in Publication Data

Bedford, A.
   Hamilton's principle in continuum mechanics.—
   (Research notes in mathematics, ISSN 0743-0337;
   v. 139)
   1. Continuum mechanics
   I. Title   II. Series
   531   QA808.2

   ISBN 0-273-08730-4

Reproduced and printed by photolithography
in Great Britain by Biddles Ltd, Guildford

Contents

Preface

The good of Hamilton is not in what he has done but in the work (not nearly half done) which he makes other people do. But to understand him you should look him up, and go through all kinds of sciences, then you go back to him, and he tells you a wrinkle.

James Clerk Maxwell

In 1808, when he was two years old, William Rowan Hamilton was sent to live with an aunt and uncle, Elizabeth and James Hamilton, in Trim, County Meath. James Hamilton was a classics scholar and graduate of Trinity College Dublin, and was headmaster of a diocesan school for boys. He soon recognized that his nephew showed extraordinary promise, and gave him intensive training in languages and the classics.

While he prepared for entrance to Trinity College, Hamilton became interested in mathematics, particularly analytic geometry. At the age of seventeen, he was reading *Théorie des Fonctions Analytiques* and *Mecanique Analytique* by Lagrange in addition to the books prescribed for the undergraduate science course at Trinity.

At Trinity College, Hamilton pursued a dual course in science and the classics—although he found it increasingly difficult to maintain his interest in the latter—and also began independent research on geometric optics as a natural extension of his interest in analytic geometry. His work led to a paper, "Theory of Systems of Rays" [38], which he presented to the Royal Irish Academy in April, 1827. Primarily on the basis of his original researches in optics, he was elected to the position of Andrews Professor of Astronomy at Trinity College in June, 1827.

Hamilton's theory of ray optics was a variational theory. It was based upon the principle, due to Fermat, that a light ray traveling between two points will follow the path which requires the least time. In the course of his work on optics, he also began to consider the possibility of developing an analogous theory for the dynamics of systems of particles. This resulted, in 1834–1835, in two papers, "On a General Method in Dynamics" [39] and "Second Essay on a General Method in

Dynamics" [40]. In the second paper he presented the result that is known today as Hamilton's principle.

Hamilton's general and elegant work on dynamics was widely quoted but not extensively applied during the remainder of the nineteenth century. However, when quantum mechanics was developed in this century, it was realized that Hamilton's work was the most natural setting for its formulation. In fact, in retrospect Hamilton's formal analogy between optics and classical mechanics was seen as a precursor of wave mechanics.

A somewhat similar historical development has occurred in the field of continuum mechanics. Although formulations of Hamilton's principle for continua began to appear as early as 1839, with the exception of applications to structural analysis, variational methods in continuum mechanics were regarded as academic since the same results could be obtained using other, more direct, methods. Some modern treatises on continuum mechanics do not mention variational methods. During the past few decades, however, interest in variational methods has increased markedly. They have been used to obtain approximate solutions, as in the finite element method, and to study the stability of solutions to problems in fluid and solid mechanics. Variational formulations have also been used to develop generalizations of the classical theories of fluid and solid mechanics.

The objective of this monograph is to give a comprehensive account of the use of Hamilton's principle to derive the equations which govern the mechanical behavior of continuous media. The classical theories of fluid and solid mechanics are discussed as well as two generalizations of those theories for which Hamilton's principle is particularly suited—materials with microstructure and mixtures.

These topics are brought together for the first time to acquaint readers who are new to this subject with an interesting and powerful alternative approach to the formulation of continuum theories. Persons interested in fluid and solid mechanics will gain a broadened perspective on those subjects as well as learn the fundamental background required to read the large literature on variational methods in continuum mechanics. For readers who are familiar with these methods, a number of recent results are presented on applications of Hamilton's principle to generalized continua and to materials containing singular surfaces. These results are presented in a setting that could encourage generalizations and extensions.

Hamilton's principle was originally expressed in terms of the classical mechanics of systems of particles. The concepts and the terminology involved in applying

Hamilton's principle to continuum mechanics are quite similar, and some familiarity with the applications to systems of particles is very helpful in understanding the extension to the case of a continuum. The application of Hamilton's principle to systems of particles is therefore discussed briefly in Chapter 1. This subject provides a simple context in which to introduce the variational ideas underlying Hamilton's principle as well as the method of Lagrange multipliers and the concept of virtual work.

Chapter 2 provides a brief survey of the mathematics and elements of continuum mechanics that are required in the following chapters. Most of this chapter can be skipped by persons who are familiar with modern continuum mechanics; however, even those who are acquainted with variational methods in continuum mechanics should briefly examine Section 2.3 before proceeding to the following chapters.

Applications of Hamilton's principle to a continuous medium are described in Chapter 3. Ideal fluids and elastic solids are treated in Subsections 3.1.1 and 3.1.2. The general case of a continuum which does not exhibit microstructural effects is presented in Subsection 3.1.3. Section 3.2 presents applications of Hamilton's principle to two particular theories of materials with microstructure. These applications illustrate the use of Hamilton's principle to generalize the ordinary theories of fluid and solid mechanics. Persons who are new to this subject may omit this section and the following chapter in a first reading.

As another example of the use of Hamilton's principle to develop generalized continuum theories, applications to mixtures are described in Chapter 4. The fact that the sum of the volume fractions of the constituents of a mixture must equal one at each point can be introduced into Hamilton's principle using the method of Lagrange multipliers. As a result of "wrinkles" such as this, Hamilton's principle provides a simple and elegant way to derive continuum theories of mixtures. A mixture of ideal fluids is discussed in Section 4.2. The case of a liquid containing a distribution of bubbles is treated as an example, including the microkinetic energy associated with bubble oscillations. In Section 4.3, a mixture of an ideal fluid and an elastic material is considered, and it is shown that the equations obtained through Hamilton's principle are equivalent to the Biot equations. A theory of mixtures of materials with microstructure in which the constituents need not be ideal or elastic is presented in Section 4.4.

In Chapter 5, a discussion is given of the application of Hamilton's principle to a continuous medium containing a surface across which the fields which characterize the medium, or their derivatives, suffer jump discontinuities. The fundamental results required to include a singular surface in a statement of Hamilton's principle are presented in Section 5.1. An elastic ideal fluid is treated as an example in Section 5.2, and it is shown that Hamilton's principle yields the jump conditions of momentum and energy across the surface.

The results presented in this monogaph are expressed in a modern framework. Persons wishing to gain an impression of Hamilton's research in its original form should consult his collected works [41],[42]. The definitive references on Hamilton's life and work are Graves [33] and Hankins [43]. In Chapters 1–3, the sources that have been used are cited, but no attempt is made to give complete or original references except for results that are relatively recent. In Chapters 2 and 3, particular reference is made to works by M. E. Gurtin. The responsibility for errors or misinterpretations of course rests with the author. Chapters 4 and 5 are based in large part on work done by the author in collaboration with D. S. Drumheller and G. Batra during the past ten years. One motivation for writing this work was to present these results in their classical context, together with a complete discussion of the foundations.

Hamilton's research anticipated modern trends in mechanics in two respects. He approached problems primarily from the perspective of a mathematician, and he consistently sought the greatest possible generality in his results. It is a measure of his success that, one hundred and fifty years after the publication of his two great works on mechanics, his results continue to find new and fruitful applications.

1 Mechanics of systems of particles

1.1 The First Problem of the Calculus of Variations

Before Hamilton's principle is introduced, some preliminary comments on the calculus of variations are necessary. Hamilton's principle is closely related to what is called the *first problem of the calculus of variations*, which can be introduced by a simple example.

Let x be a real variable, and let the closed interval $x_1 \le x \le x_2$ be denoted by $[x_1, x_2]$. A function $y(x)$ is said to be C^N on $[x_1, x_2]$ if the Nth derivative of $y(x)$ exists and is continuous on $[x_1, x_2]$. The value of a derivative at an endpoint is defined to be the limit of the derivative as the endpoint is approached from within the interval.

Let x_1, y_1 and x_2, y_2 be two fixed points in the x–y plane, with $x_1 < x_2$, and let $y(x)$ be a C^1 function on $[x_1, x_2]$ such that $y(x_1) = y_1$ and $y(x_2) = y_2$. Thus $y(x)$ describes a smooth curve which joins the two points, as shown in Figure 1.1.

The length of the curve joining the two points is

$$L = \int_{x_1}^{x_2} \left[1 + (y')^2 \right]^{\frac{1}{2}} dx, \tag{1.1}$$

where $y' = \dfrac{dy}{dx}$. Consider the following question: Can a smooth curve joining the two points be found such that its length is a minimum in comparison with other such curves? That is, among functions $y(x)$ which are C^1 on $[x_1, x_2]$ and satisfy the conditions $y(x_1) = y_1$, $y(x_2) = y_2$, can one be found for which the value of the integral (1.1) is a minimum?

The first problem of the calculus of variations is a slight generalization of this simple problem. Consider the integral

$$I = \int_{x_1}^{x_2} f(x, y, y') \, dx, \tag{1.2}$$

where f is a given function of the arguments x, y, and y', and the values x_1, x_2, $y(x_1) = y_1$ and $y(x_2) = y_2$ are prescribed. The value of the integral (1.2) depends on the function $y(x)$. A scalar-valued function such as this whose argument is itself a

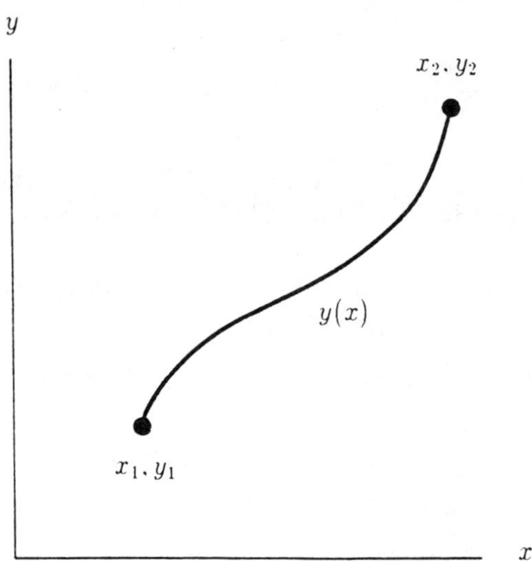

Figure 1.1: **A** smooth curve joining two points in the x–y plane.

function is called a *functionale*. As in the previous example, the question is whether a function $y(x)$ can be found such that the value of the integral is a minimum.

Certain restrictions are imposed on the functions $y(x)$ and f by the statement of the problem, the procedures that will be used in seeking its solution, and often by the physical nature of a specific application. Here consideration will be limited to functions $y(x)$ which satisfy the prescribed values at x_1 and x_2 and which are C^2 on $[x_1, x_2]$. Functions $y(x)$ having these properties will be called *admissible*. It will also be assumed that the second partial derivatives of the function f exist and are continuous on a suitable open domain of its arguments. The reason for these smoothness assumptions will become apparent.

In order to seek an admissible function $y(x)$ for which the value of the integral (1.2) is a minimum, let an admissible *comparison function* be defined by

$$y^*(x, \varepsilon) = y(x) + \varepsilon\eta(x), \tag{1.3}$$

where ε is a parameter and $\eta(x)$ is an arbitrary C^2 function on $[x_1, x_2]$ subject to the requirements that $\eta(x_1) = 0$ and $\eta(x_2) = 0$ (see Figure 1.2). If the comparison function (1.3) is substituted into the integral (1.2) in place of the function $y(x)$, the

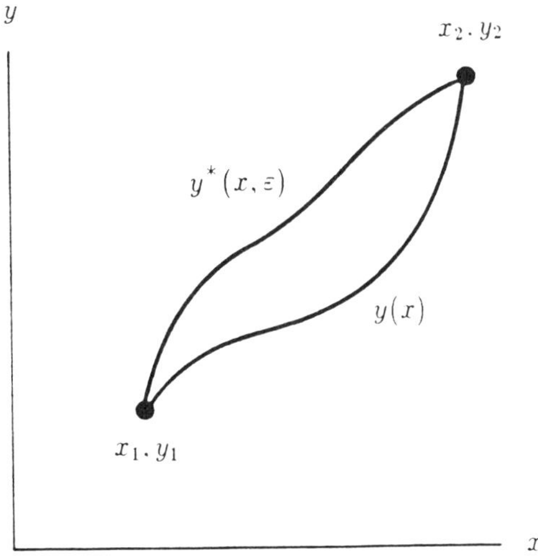

Figure 1.2: The function $y(x)$ and the comparison function $y^*(x,\varepsilon)$.

integral becomes

$$I^*(\varepsilon) = \int_{x_1}^{x_2} f\left(x, y^*, y^{*'}\right) dx, \tag{1.4}$$

where it is indicated that the value of the integral is a function of the parameter ε.

Now let it be assumed that the value of the integral (1.4) is a minimum when the comparison function $y^*(x,\varepsilon) = y(x)$. This implies that the function $I^*(\varepsilon)$ is a minimum when the parameter $\varepsilon = 0$, so that

$$\left[\frac{dI^*(\varepsilon)}{d\varepsilon}\right]_{\varepsilon=0} = 0. \tag{1.5}$$

The derivative of (1.4) with respect to ε is

$$\begin{aligned}
\frac{dI^*(\varepsilon)}{d\varepsilon} &= \int_{x_1}^{x_2} \left(\frac{\partial f^*}{\partial y^*}\frac{\partial y^*}{\partial \varepsilon} + \frac{\partial f^*}{\partial y^{*'}}\frac{\partial y^{*'}}{\partial \varepsilon}\right) dx \\
&= \int_{x_1}^{x_2} \left(\frac{\partial f^*}{\partial y^*}\eta + \frac{\partial f^*}{\partial y^{*'}}\eta'\right) dx,
\end{aligned} \tag{1.6}$$

where $f^* = f\left(x, y^*, y^{*'}\right)$ and $\eta' = \frac{d\eta}{dx}$. Therefore the condition (1.5) states that

$$\int_{x_1}^{x_2} \left(\frac{\partial f}{\partial y}\eta + \frac{\partial f}{\partial y'}\eta'\right) dx = 0. \tag{1.7}$$

The second term in this expression can be integrated by parts to obtain

$$\int_{x_1}^{x_2} \frac{\partial f}{\partial y'}\eta' dx = \left[\frac{\partial f}{\partial y'}\eta\right]_{x_1}^{x_2} - \int_{x_1}^{x_2} \frac{d}{dx}\left(\frac{\partial f}{\partial y'}\right)\eta dx. \tag{1.8}$$

Using this result and recalling that $\eta(x)$ vanishes at x_1 and at x_2, (1.7) can be written

$$\int_{x_1}^{x_2} \left[\frac{\partial f}{\partial y} - \frac{d}{dx}\left(\frac{\partial f}{\partial y'}\right)\right]\eta dx = 0. \tag{1.9}$$

Since the function $\eta(x)$ is arbitrary subject to the conditions that it be C^2 on $[x_1, x_2]$ and that it vanish at x_1 and at x_2, the expression which multiplies $\eta(x)$ in the integrand of (1.9) must vanish on $[x_1, x_2]$. If this were not the case, a function $\eta(x)$ could be chosen so that Equation (1.9) would be violated.

The formal statement of this result is called the *fundamental lemma of the calculus of variations* (see *e.g.* Bolza [11], p. 20): *Suppose that a function $\psi(x)$ is C^0 on $[x_1, x_2]$. If the equation*

$$\int_{x_1}^{x_2} \psi(x)\eta(x)dx = 0 \tag{1.10}$$

holds for every C^∞ function $\eta(x)$ on $[x_1, x_2]$ that satisfies the conditions $\eta(x_1) = 0$ and $\eta(x_2) = 0$, then $\psi(x)$ must vanish on $[x_1, x_2]$. [1]

Observe that in order to apply this lemma to Equation (1.9), the functions $y(x)$ and f must be smooth enough so that the expression which multiplies $\eta(x)$ is continuous on $[x_1, x_2]$. This is the reason for the differentiability requirements that were imposed on these functions. Note that

$$\frac{d}{dx}\left(\frac{\partial f}{\partial y'}\right) = \frac{\partial^2 f}{\partial x \partial y'} + \frac{\partial^2 f}{\partial y \partial y'}y' + \frac{\partial^2 f}{\partial y' \partial y'}y'', \tag{1.11}$$

where $y'' = \frac{d^2 y}{dx^2}$. Therefore the second derivative of $y(x)$ and the second partial derivatives of f must exist and be continuous on $[x_1, x_2]$.

On the basis of the fundamental lemma, (1.9) implies that

$$\frac{\partial f}{\partial y} - \frac{d}{dx}\left(\frac{\partial f}{\partial y'}\right) = 0 \quad \text{on } [x_1, x_2]. \tag{1.12}$$

---

[1] Proofs of several more general forms of this lemma are presented in Section 2.4.

This is called the *Euler-Lagrange equation*. It provides a differential equation with which to determine the function $y(x)$. In the case of the simple example (1.1), (1.12) yields the equation

$$y' = \text{constant}, \tag{1.13}$$

which does, of course, define the function of minimum length.

The condition (1.5) is obviously only a necessary condition, not a sufficient condition, for the value of the integral to be a minimum when $\varepsilon = 0$. This condition is also satisfied if the value of the integral is a maximum or has an inflection point of zero slope at $\varepsilon = 0$. When this condition is satisfied, the value of the integral is said to be *stationary* at $\varepsilon = 0$. Thus the condition (1.5) and the determined solution $y(x)$ are *necessary conditions given that the value of the integral is stationary in comparison with neighboring admissible functions.*

The open domain on which the second partial derivatives of the function f must be assumed to exist and be continuous can be defined in retrospect. It must encompass the values of the arguments of f associated with the solution $y(x)$ and with admissible functions (1.3) in a neighborhood of the solution.[2]

Recommended references on the calculus of variations include Akhiezer [1], Bliss [10], Bolza [11], Courant and Hilbert [15], Finlayson [28], Funk [29], Gelfand and Fomin [30], Pars [61], Washizu [73], and Weinstock [74].

1.2 Conservative Systems

1.2.1 Hamilton's Principle

Consider a system of particles or rigid bodies whose position, or *configuration*, can be described by a set of independent generalized coordinates q_k, $k = 1, 2, \ldots, K$. Let t_1 and t_2 be fixed times, with $t_1 < t_2$, and suppose that the configurations of the system at times t_1 and t_2 are prescribed. An *admissible motion* of the system will be defined to be a set of functions $q_k(t), k = 1, 2, \ldots, K$, which satisfy the prescribed values at t_1 and t_2 and which are C^2 on $[t_1, t_2]$.

Let it be assumed that the kinetic energy of the system, T, can be expressed as a function of the generalized coordinates and their time derivatives, $T = T(q_k, \dot{q}_k)$. This expression indicates that T may be a function of q_k and $\dot{q}_k$ for each value of k from 1 to K. It will also be assumed that the system is subject only to conservative

---

[2]Henceforth, when a function is said to be continuous with no additional provisos, it will be understood to be continuous on a suitable open domain of its arguments.

forces and that the potential energy of the system, U, can be expressed as a function of the generalized coordinates, $U = U(q_k)$. Each of the second partial derivatives of T and each of the first partial derivatives of U will be assumed to exist and to be continuous.[3]

What will be called the *first form* of Hamilton's principle for a system of particles states: *Among admissible motions, the actual motion of the system is such that the value of the integral*

$$I = \int_{t_1}^{t_2} (T - U)\,dt \tag{1.14}$$

is stationary in comparison with neighboring admissible motions.

Suppose that the functions $q_k(t)$ describe the actual motion of the system. In analogy with Equation (1.3), an admissible *comparison motion* of the system will be defined by

$$q_k^*(t, \varepsilon) = q_k(t) + \varepsilon \eta_k(t), \tag{1.15}$$

$k = 1, 2, \ldots, K$, where the $\eta_k(t)$ are arbitrary C^2 functions on $[t_1, t_2]$ subject to the requirements that $\eta_k(t_1) = 0$ and $\eta_k(t_2) = 0$. Upon substituting (1.15) into (1.14) in place of the functions $q_k(t)$, one obtains the integral

$$I^*(\varepsilon) = \int_{t_1}^{t_2} \left(T^* - U^* \right) dt, \tag{1.16}$$

where $T^* = T\left(q_k^*, \dot{q}_k^*\right)$ and $U^* = U\left(q_k^*\right)$.

Hamilton's principle states that the value of this integral is stationary when $q_k^*(t, \varepsilon) = q_k(t)$, which implies that

$$\left[\frac{dI^*(\varepsilon)}{d\varepsilon} \right]_{\varepsilon = 0} = 0. \tag{1.17}$$

The derivative of (1.16) with respect to ε is

$$\begin{aligned}
\frac{dI^*(\varepsilon)}{d\varepsilon} &= \int_{t_1}^{t_2} \left(\frac{\partial T^*}{\partial q_k^*} \frac{\partial q_k^*}{\partial \varepsilon} + \frac{\partial T^*}{\partial \dot{q}_k^*} \frac{\partial \dot{q}_k^*}{\partial \varepsilon} - \frac{\partial U^*}{\partial q_k^*} \frac{\partial q_k^*}{\partial \varepsilon} \right) dt \\
&= \int_{t_1}^{t_2} \left(\frac{\partial T^*}{\partial q_k^*} \eta_k + \frac{\partial T^*}{\partial \dot{q}_k^*} \dot{\eta}_k - \frac{\partial U^*}{\partial q_k^*} \eta_k \right) dt.
\end{aligned} \tag{1.18}$$

---

[3]In the simplest example, the "system" is a single particle. If there are no geometric constraints on its motion, the generalized coordinates are the three position coordinates of the particle in a suitable reference frame. The kinetic energy is $T = \frac{1}{2} m \mathbf{v} \cdot \mathbf{v}$, where m is the mass of the particle and $\mathbf{v}$ is its velocity vector. The potential energy U is defined such that $dU = -\mathbf{F} \cdot \mathbf{v}\,dt$, where $\mathbf{F}$ is the force vector acting on the particle. If such a function U exists, $\mathbf{F}$ is called *conservative*.

In this equation, use is made of the *summation convention: Whenever an index appears twice in a single expression, the expression is assumed to be summed over the range of the index.* For example,

$$\frac{\partial T^*}{\partial q_k^*}\frac{\partial q_k^*}{\partial \varepsilon} = \frac{\partial T^*}{\partial q_1^*}\frac{\partial q_1^*}{\partial \varepsilon} + \frac{\partial T^*}{\partial q_2^*}\frac{\partial q_2^*}{\partial \varepsilon} + \cdots + \frac{\partial T^*}{\partial q_K^*}\frac{\partial q_K^*}{\partial \varepsilon}. \tag{1.19}$$

This useful convention will be used throughout this work.

From Equation (1.18), the condition (1.17) is

$$\int_{t_1}^{t_2} \left(\frac{\partial T}{\partial q_k}\eta_k + \frac{\partial T}{\partial \dot{q}_k}\dot{\eta}_k - \frac{\partial U}{\partial q_k}\eta_k \right) dt = 0. \tag{1.20}$$

When the second term is integrated by parts, this equation can be written

$$\int_{t_1}^{t_2} \left[\frac{\partial L}{\partial q_k} - \frac{d}{dt}\left(\frac{\partial L}{\partial \dot{q}_k} \right) \right] \eta_k dt = 0, \tag{1.21}$$

where $L = T - U$ is the *Lagrangian* of the system. Since the functions $\eta_k(t)$ are arbitrary subject to the requirements stated above, they can be assumed to be nonzero on $[t_1, t_2]$ for $k = 1$ only. Equation (1.21) is then of the form (1.10), and the fundamental lemma applies. Repeating this process for each value of k results in the differential equations

$$\frac{\partial L}{\partial q_k} - \frac{d}{dt}\left(\frac{\partial L}{\partial \dot{q}_k} \right) = 0 \quad \text{on } [t_1, t_2] \tag{1.22}$$

for each value of k from 1 to K. These are *Lagrange's equations of motion* for the system (see *e.g.* Goldstein [31], Chapter 2).

Hamilton's principle is a postulate regarding the motion of the system. It embodies the physics of the problem. The mathematical task is to deduce the equations of motion, which are obtained as necessary conditions implied by the postulate. The number of equations of motion is equal to the number of independent generalized coordinates.

As an illustration, consider the motion of a single particle in the x–y plane. Suppose that the particle is subject only to its own weight and let the y axis be directed upward. The kinetic energy is

$$T = \tfrac{1}{2}m\left(\dot{x}^2 + \dot{y}^2 \right), \tag{1.23}$$

where m is the mass of the particle, and the potential energy is

$$U = mgy, \tag{1.24}$$

where g is the acceleration of gravity (assumed constant). Equation (1.22) yields the equations of motion

$$\begin{aligned} \ddot{x} &= 0, \\ \ddot{y} &= -g. \end{aligned} \qquad (1.25)$$

Expressions which depend on the parameter ε have been denoted by an asterisk. In applications of variational methods, the derivatives of such expressions with respect to ε, evaluated at $\varepsilon = 0$, appear frequently. This can be seen, for example, in obtaining Equation (1.20) from Equations (1.16) and (1.17). It is therefore convenient to introduce the notation[4]

$$\delta(\cdot) \equiv \left[\frac{\partial}{\partial \varepsilon}(\cdot)^* \right]_{\varepsilon = 0}. \qquad (1.26)$$

The symbol $\delta(\cdot)$ is called the *variation* of the expression $(\cdot)$.

Observe from Equation (1.15) that

$$\delta q_k = \eta_k(t). \qquad (1.27)$$

Also, from (1.16), the necessary condition (1.17) can be written

$$\int_{t_1}^{t_2} \delta(T - U)dt = 0. \qquad (1.28)$$

Stating that this equation holds for admissible comparison functions (1.15) is clearly equivalent to the first form of Hamilton's principle for a conservative system of particles. Therefore, what will be called the *second form* of Hamilton's principle for such a system states: *Among admissible motions* (1.15), *the actual motion of the system is such that* (1.28) *holds*. This is the form in which the principle was stated in Hamilton's original work [40].

1.2.2 Constraints

Thus far it has been assumed that the generalized coordinates q_k are independent. Suppose that instead they are required to satisfy prescribed equations

$$\alpha_p(q_k) = 0, \qquad (1.29)$$

---

[4]This notation, which is very common in the literature on variational methods, has acquired a bad reputation in some circles due to a history of vague definitions and a tendency to use it in performing complicated operations that are bewildering to the uninitiated. After initial attempts to write this monograph without using it, the author decided that it is too useful to discard. Throughout this work, this notation should be interpreted *only* as a symbol representing the operation (1.26).

where $p = 1, 2, \ldots, P$, $P < K$. The first partial derivatives of the functions α_p with respect to each of the q_k will be assumed to exist and to be continuous.

Hamilton's principle can be stated so that it embodies the *constraints* (1.29) by using the method of Lagrange multipliers (see *e.g.* Pars [61], Chapter VIII). The first form states: *Among admissible motions, the actual motion of the system is such that the value of the integral*

$$I = \int_{t_1}^{t_2} (T - U + C)\, dt \tag{1.30}$$

is stationary in comparison with neighboring admissible motions. The term C is defined by

$$C = \pi_p \alpha_p. \tag{1.31}$$

The unknown functions $\pi_p(t)$, $p = 1, 2, \ldots, P$, which are assumed to be C^0 on $[t_1, t_2]$, are called *Lagrange multipliers.* In determining the equations of motion, the generalized coordinates q_k may be treated as if they are *independent*; the constraints (1.29) are accounted for by introducing the Lagrange multipliers into (1.30).

Substituting the comparison motion (1.15) into (1.30) in place of the functions $q_k(t)$ yields the integral

$$I^*(\varepsilon) = \int_{t_1}^{t_2} \left(T^* - U^* + C^*\right) dt, \tag{1.32}$$

where $C^* = \pi_p(t)\alpha_p(q_k^*)$. In this case the condition (1.17) is

$$\int_{t_1}^{t_2} \left[\frac{\partial L}{\partial q_k} - \frac{d}{dt}\left(\frac{\partial L}{\partial \dot{q}_k}\right) + \pi_p \frac{\partial \alpha_p}{\partial q_k}\right] \eta_k dt = 0, \tag{1.33}$$

and the same argument used to obtain Equations (1.22) results in the differential equations of motion

$$\frac{\partial L}{\partial q_k} - \frac{d}{dt}\left(\frac{\partial L}{\partial \dot{q}_k}\right) + \pi_p \frac{\partial \alpha_p}{\partial q_k} = 0 \quad \text{on } [t_1, t_2] \tag{1.34}$$

for each value of k from 1 to K. Equations (1.29) and (1.34) provide $K + P$ equations with which to determine the generalized coordinates $q_k(t)$ and the Lagrange multipliers $\pi_p(t)$.

Returning to the example of the motion of a single particle subject to its own weight (see page 7), suppose that the particle slides without friction along a wire which constrains its motion to the path $y = x^2$. Then there is a single constraint equation

$$\alpha(x, y) = y - x^2 = 0, \tag{1.35}$$

and the equations of motion obtained from (1.34) are

$$m\ddot{x} = -2x\pi,$$

$$m\ddot{y} = \pi - mg. \tag{1.36}$$

Lagrange multipliers introduced into Hamilton's principle can be interpreted as generalized forces which cause the corresponding constraints to be satisfied. In this example, it is easy to see that the Lagrange multiplier π is the vertical component of the force exerted on the particle by the wire.

By substituting Equation (1.32) into the condition (1.17), the second form of Hamilton's principle for a conservative system of particles with constraints is obtained: *Among admissible motions (1.15), the actual motion of the system is such that*

$$\int_{t_1}^{t_2} [\delta(T - U) + \delta C]\,dt = 0. \tag{1.37}$$

1.3 Nonconservative Systems

It is a common misconception that variational methods such as Hamilton's principle are only applicable to conservative systems. Since so many interesting problems, including many problems involving continuous media, involve nonconservative forces, this would make the range of applications of Hamilton's principle very limited indeed. One objective of this monograph is to help dispel this myth.

Let the *generalized forces* Q_k be defined by

$$Q_k = -\frac{\partial U}{\partial q_k}. \tag{1.38}$$

Noting that

$$\frac{\partial U^*}{\partial \varepsilon} = \frac{\partial U^*}{\partial q_k^*}\frac{\partial q_k^*}{\partial \varepsilon} = \frac{\partial U^*}{\partial q_k^*}\eta_k \tag{1.39}$$

and using Equations (1.26), (1.27), and (1.38), one obtains

$$\delta U = -Q_k\delta q_k. \tag{1.40}$$

Using this expression, Equation (1.28) assumes the form

$$\int_{t_1}^{t_2} (\delta T + Q_k\delta q_k)\,dt = 0. \tag{1.41}$$

Of course, the system being dealt with is still a conservative one. The only thing that has been done is to introduce the notation (1.38). However, *if Hamilton's principle is postulated in terms of Equation* (1.41), *the generalized forces Q_k need not be assumed to be conservative.* Thus the form of Equation (1.41) is suggested by Hamilton's principle for a conservative system, but a new postulate is introduced in the case of a nonconservative system. The term

$$\delta W = Q_k \delta q_k \tag{1.42}$$

is called the *virtual work.*

Hamilton's principle for a nonconservative, unconstrained system of particles states: *Among admissible motions* (1.15), *the actual motion of the system is such that*

$$\int_{t_1}^{t_2} (\delta T + \delta W)\, dt = 0. \tag{1.43}$$

Clearly, if the system is conservative this postulate is identical to the statement of the second form of Hamilton's principle on page 8 . In that case, the generalized forces are derivable from the potential energy through Equation (1.38). If the system is not conservative, the generalized forces must be prescribed. Two cases occur frequently:

1. Generalized forces that are prescribed explicitly as functions of time;

2. Generalized forces that are prescribed implicitly through *constitutive equations* of the generalized coordinates and their derivatives.

Both of these cases will arise in applications of Hamilton's principle to continuous media.

A system may be subjected to both conservative and nonconservative forces, and it is often convenient to introduce the potential energy associated with the conservative forces. In this case, (1.43) is written

$$\int_{t_1}^{t_2} \left[\delta(T - U) + \delta W \right] dt = 0. \tag{1.44}$$

By using the definition (1.26), it is easy to show that

$$\delta T = \frac{\partial T}{\partial q_k} \delta q_k + \frac{\partial T}{\partial \dot{q}_k} \delta \dot{q}_k, \quad \delta U = \frac{\partial U}{\partial q_k} \delta q_k, \tag{1.45}$$

so that (1.44) can be written

$$\int_{t_1}^{t_2} \left(\frac{\partial T}{\partial q_k} \delta q_k + \frac{\partial T}{\partial \dot{q}_k} \delta \dot{q}_k - \frac{\partial U}{\partial q_k} \delta q_k + Q_k \delta q_k \right) dt = 0. \tag{1.46}$$

Integrating the second term by parts and using the fundamental lemma yields the differential equations of motion

$$\frac{d}{dt}\left(\frac{\partial L}{\partial \dot{q}_k}\right) - \frac{\partial L}{\partial q_k} = Q_k \quad \text{on } [t_1, t_2] \tag{1.47}$$

for each value of k from 1 to K. These are Lagrange's equations of motion for a system which involves both conservative and nonconservative forces (see *e.g.* Goldstein [31], Chapter 2).

The problems that will be addressed in this monograph will involve both nonconservative forces and constraints, and some of them will involve conservative forces as well. This chapter will close with a statement of Hamilton's principle for a system of particles which may exhibit each of these characteristics.

Among admissible motions (1.15), *the actual motion of the system is such that*

$$\int_{t_1}^{t_2} [\delta(T - U) + \delta W + \delta C]\,dt = 0. \tag{1.48}$$

The application of Hamilton's principle to systems of particles and rigid bodies is discussed by Goldstein [31], Hamilton [40],[42], Lanczos [50], Torby [68], Weinstock [74], and Whittaker [76].

2 Foundations of continuum mechanics

2.1 Mathematical Preliminaries

2.1.1 Inner Product Spaces

Many of the variables used in continuum mechanics obey the axioms of a finite-dimensional linear vector space with an inner product, which is simply called an *inner product space* (IPS). A result that is stated in terms of an arbitrary IPS can be applied in many contexts, achieving both generality and economy of presentation. The axioms are usually familiar to persons having technical backgrounds since they arise in the study of ordinary vector analysis. The following statement of them is paraphrased from Halmos ([37], pp. 3–14, 118–122). For the purposes of this work, scalars may be assumed to be real numbers.

A *linear vector space* $\mathcal{W}$ is a set of elements called *vectors*. An operation called *addition* is defined which associates with each pair of vectors $\mathbf{x}$ and $\mathbf{y}$ in $\mathcal{W}$ a vector $\mathbf{x} + \mathbf{y}$ in $\mathcal{W}$ such that[1]

$$\mathbf{x} + \mathbf{y} = \mathbf{y} + \mathbf{x}, \tag{2.1}$$

and for any three vectors $\mathbf{x}, \mathbf{y}, \mathbf{z}$ in $\mathcal{W}$,

$$\mathbf{x} + (\mathbf{y} + \mathbf{z}) = (\mathbf{x} + \mathbf{y}) + \mathbf{z}. \tag{2.2}$$

There is a unique vector $\mathbf{o}$ in $\mathcal{W}$ such that, for each vector $\mathbf{x}$ in $\mathcal{W}$,

$$\mathbf{x} + \mathbf{o} = \mathbf{x}. \tag{2.3}$$

For each vector $\mathbf{x}$ in $\mathcal{W}$, there is a unique vector $-\mathbf{x}$ such that

$$\mathbf{x} + (-\mathbf{x}) = \mathbf{o}. \tag{2.4}$$

An operation called *scalar multiplication* is defined which associates with each scalar α and each vector $\mathbf{x}$ in $\mathcal{W}$ a vector $\alpha\mathbf{x}$ in $\mathcal{W}$ such that, for any scalars α, β and vectors

---

[1]Linear vector spaces will be denoted by script capital letters. Vectors will be denoted by bold-face letters, usually lower case, although there will be exceptions which will be defined individually.

13

$\mathbf{x}, \mathbf{y}$ in $\mathcal{W}$,

$$\alpha(\beta \mathbf{x}) = (\alpha \beta)\mathbf{x},$$

$$1\mathbf{x} = \mathbf{x},$$

$$\alpha(\mathbf{x} + \mathbf{y}) = \alpha \mathbf{x} + \alpha \mathbf{y}, \tag{2.5}$$

$$(\alpha + \beta)\mathbf{x} = \alpha \mathbf{x} + \beta \mathbf{x}.$$

A finite set of vectors $\{\mathbf{x}_k\} = \mathbf{x}_1, \mathbf{x}_2, \ldots, \mathbf{x}_N$ in $\mathcal{W}$ is called *linearly independent* if the equation

$$\alpha_1 \mathbf{x}_1 + \alpha_2 \mathbf{x}_2 + \cdots + \alpha_N \mathbf{x}_N = \alpha_k \mathbf{x}_k = \mathbf{o} \tag{2.6}$$

holds only when $\alpha_k = 0$ for each value of k from 1 to N. If such a set of vectors exists for which each vector $\mathbf{x}$ in $\mathcal{W}$ can be written in the form

$$\mathbf{x} = \beta_k \mathbf{x}_k, \tag{2.7}$$

then $\mathcal{W}$ is said to be of *dimension* N, and $\{\mathbf{x}_k\}$ is called a *basis* for $\mathcal{W}$.

The axioms and definitions stated thus far characterize a finite-dimensional linear vector space. An IPS is obtained by appending an operation called the *inner product* which associates with each pair of vectors $\mathbf{x}$ and $\mathbf{y}$ in $\mathcal{W}$ a scalar denoted by $\mathbf{x} \cdot \mathbf{y}$ such that, for any scalars α, β and vectors $\mathbf{x}, \mathbf{y}, \mathbf{z}$ in $\mathcal{W}$,

$$\mathbf{x} \cdot \mathbf{y} = \mathbf{y} \cdot \mathbf{x},$$

$$\mathbf{x} \cdot \mathbf{x} \geq 0, \tag{2.8}$$

where $\mathbf{x} \cdot \mathbf{x} = 0$ if and only if $\mathbf{x} = \mathbf{o}$, and

$$(\alpha \mathbf{x} + \beta \mathbf{y}) \cdot \mathbf{z} = \alpha(\mathbf{x} \cdot \mathbf{z}) + \beta(\mathbf{y} \cdot \mathbf{z}). \tag{2.9}$$

The *magnitude*, or *norm*, of a vector $\mathbf{x}$ is defined to be the scalar

$$|\mathbf{x}| = (\mathbf{x} \cdot \mathbf{x})^{\frac{1}{2}}. \tag{2.10}$$

The real numbers are an IPS if the inner product is defined to be the usual product of two numbers. It is one dimensional, and any number other than zero is a basis. As a second example, the three dimensional vectors of ordinary vector analysis constitute an IPS, with the usual definition of the inner product. *The symbol $\mathcal{V}$ will be reserved for this particular* IPS. A third example of an IPS which is particularly important in continuum mechanics is the set of linear transformations of $\mathcal{V}$ into $\mathcal{V}$, which will be discussed in the next subsection.

14

2.1.2 Linear Transformations

Let $\mathcal{U}$ and $\mathcal{W}$ be inner product spaces. A *linear transformation*[2] of $\mathcal{U}$ into $\mathcal{W}$, denoted by $\mathbf{L} : \mathcal{U} \to \mathcal{W}$, associates with each vector $\mathbf{u}$ in $\mathcal{U}$ a vector $\mathbf{Lu}$ in $\mathcal{W}$ such that, for any scalars α, β and vectors $\mathbf{u}$, $\mathbf{v}$ in $\mathcal{U}$,

$$\mathbf{L}(\alpha\mathbf{u} + \beta\mathbf{v}) = \alpha\mathbf{Lu} + \beta\mathbf{Lv}. \tag{2.11}$$

The sum of two linear transformations and the product of a scalar and a linear transformation are defined such that, for each scalar α and vector $\mathbf{u}$ in $\mathcal{U}$,

$$(\mathbf{L}_1 + \mathbf{L}_2)\mathbf{u} = \mathbf{L}_1\mathbf{u} + \mathbf{L}_2\mathbf{u},$$
$$(\alpha\mathbf{L})\mathbf{u} = \mathbf{L}(\alpha\mathbf{u}). \tag{2.12}$$

Recall that $\mathcal{V}$ denotes the IPS of ordinary, three-dimensional vector analysis, and consider linear transformations of $\mathcal{V}$ into itself. The rest of this subsection will be concerned with linear transformations of this type, which are called *second-order tensors*. Three simple examples are the zero tensor $\mathbf{0}$, the identity tensor $\mathbf{1}$, and the tensor product $\mathbf{u} \otimes \mathbf{v}$, which are defined such that, for each $\mathbf{u}$, $\mathbf{v}$, $\mathbf{w}$ in $\mathcal{V}$,

$$\mathbf{0}\mathbf{v} = \mathbf{o},$$
$$\mathbf{1}\mathbf{v} = \mathbf{v}, \tag{2.13}$$
$$(\mathbf{u} \otimes \mathbf{v})\mathbf{w} = \mathbf{u}(\mathbf{v} \cdot \mathbf{w}).$$

Let $\{\mathbf{e}_k\} = \mathbf{e}_1, \mathbf{e}_2, \mathbf{e}_3$ be an orthonormal basis for $\mathcal{V}$. Then each vector $\mathbf{v}$ in $\mathcal{V}$ can be written as the linear combination

$$\mathbf{v} = v_k\mathbf{e}_k, \tag{2.14}$$

where the coefficients v_k are called the components of $\mathbf{v}$ with respect to $\{\mathbf{e}_k\}$. If $\mathbf{T}$ is a linear transformation, the equation

$$\mathbf{Tu} = \mathbf{v} \tag{2.15}$$

can be written

$$\mathbf{T}u_k\mathbf{e}_k = v_k\mathbf{e}_k. \tag{2.16}$$

---

[2]Linear transformations will be denoted by bold capital letters, with exceptions which will be defined individually.

Taking the inner product of this equation with $\mathbf{e}_m$ results in the equation

$$T_{mk}u_k = v_m, \tag{2.17}$$

where the scalars

$$T_{mk} = (\mathbf{T})_{mk} = \mathbf{e}_m \cdot \mathbf{T}\mathbf{e}_k \tag{2.18}$$

are called the components of $\mathbf{T}$ with respect to $\{\mathbf{e}_k\}$. For example, the components of the linear transformations $\mathbf{u} \otimes \mathbf{v}$ and $\mathbf{1}$ with respect to $\{\mathbf{e}_k\}$ are easily shown to be

$$(\mathbf{u} \otimes \mathbf{v})_{mk} = u_m v_k, \quad (\mathbf{1})_{mk} = \delta_{mk}, \tag{2.19}$$

where the *Kronecker delta* δ_{mk} is defined by

$$\delta_{mk} = \begin{cases} 1 \text{ if } m = k, \\ 0 \text{ if } m \neq k. \end{cases} \tag{2.20}$$

The *transpose* of $\mathbf{T}$ is defined to be the linear transformation $\mathbf{T}^t$ which has the property that, for any vectors $\mathbf{u}, \mathbf{v}$ in $\mathcal{V}$,

$$\mathbf{u} \cdot \mathbf{T}\mathbf{v} = \mathbf{T}^t\mathbf{u} \cdot \mathbf{v}. \tag{2.21}$$

The components of $\mathbf{T}^t$ are

$$T^t_{km} = T_{mk}. \tag{2.22}$$

The *composition*, or *product*, of two linear transformations $\mathbf{S}$ and $\mathbf{T}$, denoted by $\mathbf{ST}$, is defined to be the linear transformation

$$\mathbf{ST}\mathbf{v} = \mathbf{S}(\mathbf{T}\mathbf{v}). \tag{2.23}$$

The components of $\mathbf{ST}$ are easily shown to be

$$(\mathbf{ST})_{km} = S_{kj}T_{jm}. \tag{2.24}$$

The *determinant* of a linear transformation $\mathbf{T}$, denoted by $\det \mathbf{T}$, is defined such that

$$\det \mathbf{T} = \det [T_{km}], \tag{2.25}$$

where $[T_{km}]$ denotes the matrix of the components of $\mathbf{T}$. Two results concerning determinants that will be useful are

$$\frac{\partial(\det \mathbf{T})}{\partial T_{km}} = \operatorname{cof} T_{km}, \quad \delta_{km} \det \mathbf{T} = T_{kj} \operatorname{cof} T_{mj}, \tag{2.26}$$

where cof T_{km} is the cofactor of the element T_{km} of $[T_{km}]$.

The *inverse* of $\mathbf{T}$ is the linear transformation $\mathbf{T}^{-1}$ such that

$$\mathbf{T}\mathbf{T}^{-1} = \mathbf{T}^{-1}\mathbf{T} = \mathbf{1}. \tag{2.27}$$

The components of $\mathbf{T}^{-1}$ are

$$T_{km}^{-1} = \ \text{cof} \ T_{mk}/\det \mathbf{T}. \tag{2.28}$$

The *trace* of $\mathbf{T}$ is defined by[3]

$$\text{tr} \ \mathbf{T} = T_{kk}, \tag{2.29}$$

and the *inner product* of two linear transformations $\mathbf{T}$ and $\mathbf{S}$ is defined by

$$\mathbf{S} \cdot \mathbf{T} = \text{tr}\left(\mathbf{S}^t\mathbf{T}\right) = S_{km}T_{km}. \tag{2.30}$$

It is easy to demonstrate that *the set of all linear transformations* $\mathbf{T} : \mathcal{V} \to \mathcal{V}$ *with the inner product* (2.30) *is an inner product space.*

2.1.3 Functions, Continuity and Differentiability

Let $\mathcal{U}$ and $\mathcal{W}$ be inner product spaces, and let U be a subset of $\mathcal{U}$. A *function* $\mathcal{F} : U \to \mathcal{W}$ associates with each vector $\mathbf{u}$ in U a vector $\mathcal{F}(\mathbf{u})$ in $\mathcal{W}$. The concept of the magnitude of a vector in an IPS, defined by Equation (2.10), makes it possible to define the limit, continuity and differentiability of the function $\mathcal{F}(\mathbf{u})$ in a manner entirely analogous to ordinary calculus.

A vector $\mathbf{w}$ in $\mathcal{W}$ is said to be the *limit* of $\mathcal{F}(\mathbf{u})$ at a vector $\mathbf{u}_0$ in $\mathcal{U}$ if for any positive scalar α there is a positive scalar β such that $|\mathcal{F}(\mathbf{u}) - \mathbf{w}| < \alpha$ for each vector $\mathbf{u}$ in U which satisfies the relation $0 < |\mathbf{u} - \mathbf{u}_0| < \beta$. The function $\mathcal{F}(\mathbf{u})$ is said to be *continuous* at a vector $\mathbf{u}_0$ in U if the limit $\mathbf{w}$ exists and $\mathcal{F}(\mathbf{u}_0) = \mathbf{w}$, and it is said to be *continuous in* U if it is continuous at each vector in U.

The set U is called an *open* subset of $\mathcal{U}$ if for each vector $\mathbf{u}_0$ in U there is a positive scalar α such that the vector $\mathbf{u}_0 + \mathbf{u}$ is in U for each vector $\mathbf{u}$ in U which satisfies $|\mathbf{u}| < \alpha$.

Let U be an open subset of $\mathcal{U}$. A function $\mathcal{F} : U \to \mathcal{W}$ is said to be *differentiable* at a vector $\mathbf{u}_0$ in U if there is a linear transformation, denoted by $\dfrac{d\mathcal{F}}{d\mathbf{u}} : \mathcal{U} \to \mathcal{W}$ such that

$$\mathcal{F}(\mathbf{u}_0) - \mathcal{F}(\mathbf{u}) = \frac{d\mathcal{F}}{d\mathbf{u}}(\mathbf{u}_0 - \mathbf{u}) + o\left(|\mathbf{u}_0 - \mathbf{u}|\right). \tag{2.31}$$

---

[3]The determinant and trace of a linear transformation can be defined in a way that is independent of any basis (see *e.g.* Bowen and Wang [13], Section 40.). The definitions given here are adequate for the purposes of this monograph.

The notation $o(\alpha)$ means that $|o(\alpha)/\alpha| \to 0$ as $\alpha \to 0$. The linear transformation $\frac{d\mathcal{F}}{d\mathbf{u}}$ is called the *derivative*[4] of $\mathcal{F}(\mathbf{u})$ at $\mathbf{u}_0$. The function $\mathcal{F}(\mathbf{u})$ is said to be *differentiable in U* if it is differentiable at each vector in U and $\frac{d\mathcal{F}}{d\mathbf{u}}$ is continuous in U.

2.1.4 Fields and the Divergence Theorem

In continuum mechanics, the properties of materials are described in terms of piece-wise continuous functions, or *fields*. Hamilton's principle for a continuous medium will be stated in terms of a prescribed volume of material. In this subsection, definitions and terminology associated with fields and volumes will be introduced.

Let a reference point O and an orthonormal basis $\{\mathbf{e}_k\}$ define an inertial reference frame in three dimensional Euclidean space $\mathcal{E}$, and let the vector $\mathbf{X}$ in $\mathcal{V}$ denote the position vector of a point in $\mathcal{E}$ relative to O. Consider a closed surface ∂B in $\mathcal{E}$. Let B be the interior of ∂B, and let the interior together with the surface (the *closure*) be denoted by $\overline{B}$.

It will be assumed that B is a *bounded regular region*, and that the surface ∂B may consist of complementary *regular subsurfaces* ∂B_1 and ∂B_2. Precise definitions of bounded regular regions and regular subsurfaces (which insure, for example, that the divergence theorem can be applied) are given by Gurtin ([35], pp. 12–14). A volume which is bounded by a single closed surface consisting of a finite number of smooth subsurfaces, each of which is bounded by a piecewise smooth curve, is a bounded regular region. If the surface of such a volume is divided into two parts by a single piecewise smooth closed curve, the resulting complementary subsurfaces are regular subsurfaces.

Let $\mathcal{W}$ be an inner product space. A *field* $\mathbf{f} : B \to \mathcal{W}$ is a function which associates with each point in B (identified with its position vector $\mathbf{X}$) a vector $\mathbf{f}(\mathbf{X})$ in $\mathcal{W}$. In the cases where the elements of $\mathcal{W}$ are scalars, ordinary vectors, or second-order tensors, $\mathbf{f}(\mathbf{X})$ is called a scalar, vector, or tensor field.

As an example, consider a scalar field $\phi(\mathbf{X})$, and let $\mathbf{Z}$ be any vector in $\mathcal{V}$. If $\phi(\mathbf{X})$ is differentiable at a point $\mathbf{X}$ in B, then

$$\frac{d\phi}{d\mathbf{X}}\mathbf{Z} = \mathrm{GRAD}\,\phi \cdot \mathbf{Z}, \qquad (2.32)$$

---

[4]A variety of notations are used for this linear transformation, including $\Delta\mathcal{F}$ and $D\mathcal{F}(\mathbf{u})$. The notation used here was chosen so that it would look familiar to persons used to ordinary derivatives, and also because it makes expressions in which the chain rule is used look much more intelligible.

18

where GRAD ϕ is the familiar gradient

$$\text{GRAD } \phi = \frac{\partial \phi}{\partial X_k} \mathbf{e}_k. \tag{2.33}$$

In the case of a vector field $\mathbf{v}(\mathbf{X})$ which is differentiable at a point $\mathbf{X}$ in B, the linear transformation $\frac{d\mathbf{v}}{d\mathbf{X}}$ is called the gradient of the vector field. In terms of components,

$$\left(\frac{d\mathbf{v}}{d\mathbf{X}} \right)_{km} = \frac{\partial v_k}{\partial X_m}. \tag{2.34}$$

Note that the divergence of $\mathbf{v}(\mathbf{X})$ is

$$\text{DIV } \mathbf{v} = \text{tr } \frac{d\mathbf{v}}{d\mathbf{X}} = \frac{\partial v_k}{\partial X_k}. \tag{2.35}$$

The divergence of a tensor field $\mathbf{T}(\mathbf{X})$ which is differentiable at a point $\mathbf{X}$ in B is defined to be the vector DIV $\mathbf{T}$ which has the property that, for each vector $\mathbf{Z}$ in $\mathcal{V}$,

$$\text{DIV } \mathbf{T} \cdot \mathbf{Z} = \text{DIV } \left(\mathbf{T}^t \mathbf{Z} \right). \tag{2.36}$$

The components of DIV $\mathbf{T}$ are

$$(\text{DIV } \mathbf{T})_k = \frac{\partial T_{km}}{\partial X_m}. \tag{2.37}$$

Let $\mathbf{f}(\mathbf{X})$ be a field that is continuous in B, and let $\mathbf{X}_0$ be a point of ∂B. If the limit of $\mathbf{f}(\mathbf{X})$ as $\mathbf{X} \to \mathbf{X}_0$ exists at each point of ∂B and is continuous on ∂B, then $\mathbf{f}(\mathbf{X})$ is said to have a *continuous extension* on $\overline{B}$ if its value at each point $\mathbf{X}_0$ of ∂B is defined to be the value of its limit at that point.

The fields to be considered in this work will usually be functions of both position and time. A *time dependent field* $\mathbf{f} : B \times (t_1, t_2) \to \mathcal{W}$ is a function which associates with each point in B and each time in the open interval $t_1 < t < t_2$ a vector $\mathbf{f}(\mathbf{X}, t)$ in $\mathcal{W}$.

A vector $\mathbf{w}$ in $\mathcal{W}$ is said to be the *limit* of $\mathbf{f}(\mathbf{X}, t)$ at $\mathbf{X}_0, t_0$ in $B \times (t_1, t_2)$ if for any positive scalar α there is a positive scalar β such that

$$|\mathbf{f}(\mathbf{X}, t) - \mathbf{w}| < \alpha$$

for each $\mathbf{X}, t$ in $B \times (t_1, t_2)$ which satisfy the relation

$$0 < \left[|\mathbf{X} - \mathbf{X}_0|^2 + (t - t_0)^2 \right]^{\frac{1}{2}} < \beta.$$

The field $\mathbf{f}(\mathbf{X}, t)$ is said to be *continuous* at $\mathbf{X}_0, t_0$ if the limit $\mathbf{w}$ exists and $\mathbf{f}(\mathbf{X}_0, t_0) = \mathbf{w}$, and it is said to be *continuous in* $B \times (t_1, t_2)$ if it is continuous at each $\mathbf{X}_0, t_0$ in $B \times (t_1, t_2)$.

Let $\dfrac{\partial^n \mathbf{f}}{\partial \mathbf{X}^n}$ denote the nth derivative of $\mathbf{f}(\mathbf{X}, t)$ holding t fixed. Then $\mathbf{f}(\mathbf{X}, t)$ is said to be C^N in $B \times (t_1, t_2)$ if it is continuous in $B \times (t_1, t_2)$ and the derivatives

$$\frac{\partial^m}{\partial t^m} \frac{\partial^n \mathbf{f}}{\partial \mathbf{X}^n}, \quad 0 \leq m \leq N, \ \ 0 \leq n \leq N, \ \ m + n \leq N \tag{2.38}$$

exist and are continuous in $B \times (t_1, t_2)$. Such a field is then said to be C^N on $\overline{B} \times [t_1, t_2]$ if these derivatives have continuous extensions to $\overline{B} \times [t_1, t_2]$.

Let ∂B_α be a complementary regular subsurface of B, and let the vector function $\mathbf{N}(\mathbf{X})$ defined on ∂B_α be the outward directed unit vector normal to ∂B_α at each point $\mathbf{X}$ of ∂B_α. A point $\mathbf{X}$ at which $\mathbf{N}(\mathbf{X})$ is continuous is called a *regular point* of ∂B_α.

A function $\mathbf{f}(\mathbf{X})$ defined on ∂B_α is called *piecewise regular* if it is piecewise continuous on ∂B_α and is continuous at each regular point of ∂B_α. A time dependent function $\mathbf{f}(\mathbf{X}, t)$ defined on $\partial B_\alpha \times [t_1, t_2]$ is called piecewise regular if it is piecewise continuous on $\partial B_\alpha \times [t_1, t_2]$ and $\mathbf{f}(\mathbf{X}, t_0)$ is piecewise regular on ∂B_α for each fixed time t_0 on $[t_1, t_2]$. A function $\mathbf{f}(\mathbf{X}, t)$ defined on $\partial B_\alpha \times [t_1, t_2]$ is said to be *continuous in time* if for each fixed point $\mathbf{X}$ of ∂B_α it is a continuous function of time on $[t_1, t_2]$.

Two functions $\mathbf{f}_1(\mathbf{X}, t)$ and $\mathbf{f}_2(\mathbf{X}, t)$ defined on $\partial B_\alpha \times [t_1, t_2]$ are defined to be equal if for each time t on $[t_1, t_2]$ they are equal at each regular point of ∂B_α.

The divergence theorem will be used repeatedly in applying Hamilton's principle to continuous media. The following statement is paraphrased from Gurtin ([35], p. 16). Let $\phi(\mathbf{X})$, $\mathbf{v}(\mathbf{X})$, and $\mathbf{T}(\mathbf{X})$ be scalar, vector, and tensor fields which are continuous on $\overline{B}$ and differentiable in B. Then

$$\int_{\partial B} \phi \mathbf{N} dS = \int_B \mathrm{GRAD}\ \phi\, dV,$$

$$\int_{\partial B} \mathbf{v} \cdot \mathbf{N} dS = \int_B \mathrm{DIV}\ \mathbf{v}\, dV, \tag{2.39}$$

$$\int_{\partial B} \mathbf{T} \mathbf{N} dS = \int_B \mathrm{DIV}\ \mathbf{T}\, dV$$

when the integrands on the right are piecewise continuous on $\overline{B}$. Recall that $\mathbf{N}$ is the outward directed unit vector that is normal to ∂B.

Suggested references on the mathematical foundations of continuum mechanics include Bowen and Wang [13],[14], Ericksen [25], Gurtin [36], Halmos [37], Leigh [52], Truesdell and Noll [70], and Truesdell and Toupin [71].

20

2.2   Motion and Deformation

A *motion* of a material which is modeled as a continuous medium is described by a time dependent vector field

$$\mathbf{x} = \chi(\mathbf{X}, t), \tag{2.40}$$

where $\mathbf{x}$ is the position vector at time t of the *material point* identified with its position vector $\mathbf{X}$ in a reference state, or *reference configuration*. As a simple example, consider a quantity of a malleable material, such as dough, which is at rest. This rest state can be used as the reference configuration. Imagine that a point on the surface or within the material is marked with a pen. Let its position vector be $\mathbf{X}_0$. Then if the material is picked up and deformed, and equation (2.40) describes its motion, the trajectory of the marked point in space is given by

$$\mathbf{x} = \chi\left(\mathbf{X}_0, t\right). \tag{2.41}$$

Thus (2.40) describes the motion of each point of the material.

In general, it is not necessary that the reference configuration be one which the material has actually assumed at any time. However, this distinction is not needed for any of the applications to be considered in this work. The reference configuration will be assumed to be the configuration of the material at time t_1. That is,

$$\mathbf{X} = \chi\left(\mathbf{X}, t_1\right). \tag{2.42}$$

Suppose that in its reference configuration the material occupies a bounded regular region B with surface ∂B. The motion (2.40) maps the material onto a volume B_t with surface ∂B_t at time t (Figure 2.1). In keeping with the interpretation of (2.40) as the motion of a material, the mapping of the material points from $\overline{B}$ to $\overline{B}_t$ will be assumed to be *one-one*. That is, if $\mathbf{X}_1$ and $\mathbf{X}_2$ are distinct points of $\overline{B}$, then $\mathbf{x}_1 = \chi(\mathbf{X}_1, t)$ and $\mathbf{x}_2 = \chi(\mathbf{X}_2, t)$ are distinct points of $\overline{B}_t$, and for each point $\mathbf{x}$ of $\overline{B}_t$, there is a point $\mathbf{X}$ of $\overline{B}$ such that $\mathbf{x} = \chi(\mathbf{X}, t)$. This requirement insures that the *inverse motion*

$$\mathbf{X} = \chi^{-1}(\mathbf{x}, t), \tag{2.43}$$

which maps the material points from $\overline{B}_t$ onto $\overline{B}$ at time t, exists and is one-one.

Suppose that the field (2.40) is C^N on $\overline{B} \times [t_1, t_2]$, $N \geq 1$. The *deformation*

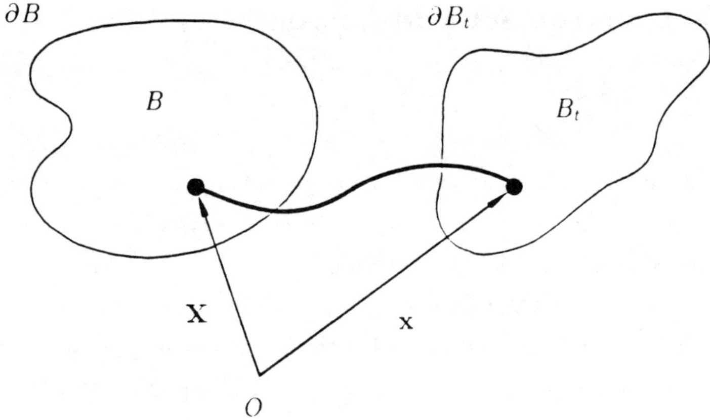

Figure 2.1: The reference configuration and the configuration at time t.

gradient **F** is the tensor field[5]

$$\mathbf{F} = \frac{\partial \chi}{\partial \mathbf{X}}, \quad F_{km} = \frac{\partial \chi_k}{\partial X_m}. \tag{2.44}$$

The *Jacobian* of the motion is defined by

$$J = \det \mathbf{F}. \tag{2.45}$$

A necessary condition for (2.40) to describe the motion of a material is that $J > 0$ in B. It will be seen that this condition insures that the volume of any element of the material remains positive. Given that this condition is satisfied, it can be shown (see *e.g.* Gurtin [36], pp. 60, 65–66) that the inverse motion (2.43) is C^N on $\overline{B}_t \times [t_1, t_2]$.

The interpretation of the motion (2.41) as describing the trajectory of a material point in space motivates the definitions of the *velocity*

$$\mathbf{v} = \frac{\partial}{\partial t} \chi(\mathbf{X}, t) \tag{2.46}$$

and the *acceleration*

$$\mathbf{a} = \frac{\partial^2}{\partial t^2} \chi(\mathbf{X}, t). \tag{2.47}$$

---

[5]Some expressions will be presented both in direct notation and in terms of components for the sake of clarity.

The inverse motion (2.43) can be used to express the velocity and acceleration as functions of $\mathbf{x}, t$. When the functional dependence of a field is not obvious from the context, a caret will be used to indicate that it is expressed in terms of $\mathbf{X}, t$. The caret will not be used when the functional dependence is shown explicitly. For example

$$\mathbf{v}(\mathbf{X}, t) = \hat{\mathbf{v}}\left(\chi^{-1}(\mathbf{x}, t), t\right) = \mathbf{v}(\mathbf{x}, t),$$

$$(2.48)$$

$$\mathbf{a} = \frac{\partial}{\partial t}\hat{\mathbf{v}} = \frac{\partial}{\partial t}\mathbf{v} + \mathbf{L}\mathbf{v},$$

where the linear transformation $\mathbf{L}$ is the *velocity gradient*

$$\mathbf{L} = \frac{\partial \mathbf{v}}{\partial \mathbf{x}}, \quad L_{km} = \frac{\partial v_k}{\partial x_m}.$$

$$(2.49)$$

The *material derivative* of a field $\mathbf{f}(\mathbf{X}, t)$ is defined by

$$\dot{\mathbf{f}} = \frac{\partial}{\partial t}\hat{\mathbf{f}}.$$

$$(2.50)$$

Thus the material derivative is the time rate of change of a field holding the material point fixed. For example, note that $\mathbf{a} = \dot{\mathbf{v}}$. In the case of a scalar field $\phi(\mathbf{X}, t)$,

$$\dot{\phi} = \frac{\partial}{\partial t}\hat{\phi} = \frac{\partial}{\partial t}\phi + \mathbf{v} \cdot \operatorname{grad} \phi,$$

$$(2.51)$$

where $\operatorname{grad} \phi = \frac{\partial \phi}{\partial x_k}\mathbf{e}_k$.

The motion (2.40) maps a volume element dV of B onto a volume element dV_t of B_t at time t. It is easy to show (see *e.g.* Truesdell and Toupin [71], pp. 247–249) that

$$dV_t = J dV.$$

$$(2.52)$$

The *density* ρ is a scalar field defined such that the mass of each volume element dV_t of B_t is ρdV_t. Let the value of ρ at time t_1 be denoted by ρ_R. That is, ρ_R is the density of the reference configuration. Then one form of the equation of *conservation of mass* is

$$\rho dV_t = \rho_R dV.$$

$$(2.53)$$

Using Equation (2.52), this relation can be expressed in the form

$$J = \rho_R / \rho.$$

$$(2.54)$$

The material derivative of J is

$$\dot{J} = \frac{\partial(\det \mathbf{F})}{\partial F_{km}}\frac{\partial \hat{F}_{km}}{\partial t},$$

$$(2.55)$$

and from (2.44),

$$\frac{\partial \hat{F}_{km}}{\partial t} = \frac{\partial^2 \chi_k}{\partial t \partial X_m} = \frac{\partial v_k}{\partial x_p}\frac{\partial x_p}{\partial X_m} = L_{kp}F_{pm}. \tag{2.56}$$

Substituting this result into (2.55) and using (2.26) yields the relation

$$\dot{J} = J \operatorname{tr} \mathbf{L} = J \operatorname{div} \mathbf{v}, \tag{2.57}$$

where $\operatorname{div} \mathbf{v} = \dfrac{\partial v_k}{\partial x_k}$. Taking the material derivative of (2.54) and using Equation (2.57) leads to the equation of conservation of mass in its more familiar form

$$\dot{\rho} + \rho \operatorname{div} \mathbf{v} = 0. \tag{2.58}$$

The motion (2.40) maps a surface element dS of ∂B onto a surface element dS_t of ∂B_t at time t. Let the function $\mathbf{n}(\mathbf{x},t)$ defined on ∂B_t denote the outward directed unit vector that is normal to ∂B_t, and let $\mathbf{N}(\mathbf{X}) = \hat{\mathbf{n}}(\mathbf{X},t_1)$. That is, $\mathbf{N}$ is the outward directed unit vector normal to ∂B. It can be shown (see *e.g.* Truesdell and Toupin [71], pp. 247–249) that

$$\mathbf{n}dS_t = J\mathbf{F}^{-t}\mathbf{N}dS, \tag{2.59}$$

where $\mathbf{F}^{-t} = \left(\mathbf{F}^{-1}\right)^{t}$.

By means of the relations (2.52) and (2.59), integrals on B and ∂B can be expressed as integrals on B_t and ∂B_t, and *vice versa*. If a field $\mathbf{f}(\mathbf{X},t)$ is continuous on $\overline{B} \times [t_1,t_2]$, then[6]

$$\int_{B_t} \mathbf{f}\,dV_t = \int_{B} \mathbf{f}\,J\,dV. \tag{2.60}$$

Similarly, if a scalar function $\phi(\mathbf{X},t)$ defined on ∂B is piecewise regular, then

$$\int_{\partial B_t} \phi\mathbf{n}\,dS_t = \int_{\partial B} \phi J\mathbf{F}^{-t}\mathbf{N}\,dS. \tag{2.61}$$

Consider two neighboring material points in B having position vectors $\mathbf{X}$ and $\mathbf{X} + d\mathbf{X}$. The square of the distance separating them is

$$dS^2 = d\mathbf{X} \cdot d\mathbf{X}. \tag{2.62}$$

---

[6]In Equation (2.60) the functional dependence of the field $\mathbf{f}$ is indicated by the context. It must be expressed in terms of $\mathbf{x},t$ in the left integral, and in terms of $\mathbf{X},t$ in the right integral.

At time t, the same two material points are separated by the vector

$$d\mathbf{x} = \chi_k\left(X_m + dX_m, t\right)\mathbf{e}_k - \chi_k\left(X_m, t\right)\mathbf{e}_k$$

$$= \frac{\partial \chi_k}{\partial X_m} dX_m \mathbf{e}_k \tag{2.63}$$

$$= \mathbf{F}\, d\mathbf{X},$$

so that the square of the distance separating them at time t is

$$ds^2 = d\mathbf{x} \cdot d\mathbf{x} = d\mathbf{X} \cdot \mathbf{F}^t \mathbf{F}\, d\mathbf{X}. \tag{2.64}$$

Therefore

$$ds^2 - dS^2 = d\mathbf{X} \cdot (\mathbf{C} - \mathbf{1})d\mathbf{X}, \tag{2.65}$$

where

$$\mathbf{C} = \mathbf{F}^t \mathbf{F} \tag{2.66}$$

is called the *right Cauchy-Green strain tensor.* Since Equation (2.65) determines the change in the distance between any two neighboring material points at time t, the deformation gradient $\mathbf{F}$ (or deformation measures expressed in terms of $\mathbf{F}$) *determines the deformation of the material in the neighborhood of a material point.*

The *displacement* is the vector field

$$\mathbf{u} = \chi(\mathbf{X}, t) - \mathbf{X}. \tag{2.67}$$

It is the displacement vector of a material point relative to its position in the reference configuration. The *displacement gradient* is the tensor field

$$\frac{\partial \mathbf{u}}{\partial \mathbf{X}} = \mathbf{F} - \mathbf{1}, \quad \frac{\partial u_k}{\partial X_m} = F_{km} - \delta_{km}. \tag{2.68}$$

In terms of the displacement gradient,

$$\mathbf{C} - \mathbf{1} = 2\mathbf{E} + \left(\frac{\partial \mathbf{u}}{\partial \mathbf{X}}\right)^t \frac{\partial \mathbf{u}}{\partial \mathbf{X}}, \tag{2.69}$$

where $\mathbf{E}$ is the *linear strain*

$$\mathbf{E} = \tfrac{1}{2}\left[\frac{\partial \mathbf{u}}{\partial \mathbf{X}} + \left(\frac{\partial \mathbf{u}}{\partial \mathbf{X}}\right)^t\right], \quad E_{km} = \tfrac{1}{2}\left(\frac{\partial u_k}{\partial X_m} + \frac{\partial u_m}{\partial X_k}\right). \tag{2.70}$$

Recommended references on the motion and deformation of a continuous medium include Eringen [27], Gurtin [36], Leigh [52], Truesdell and Noll [70], and Truesdell and Toupin [71].

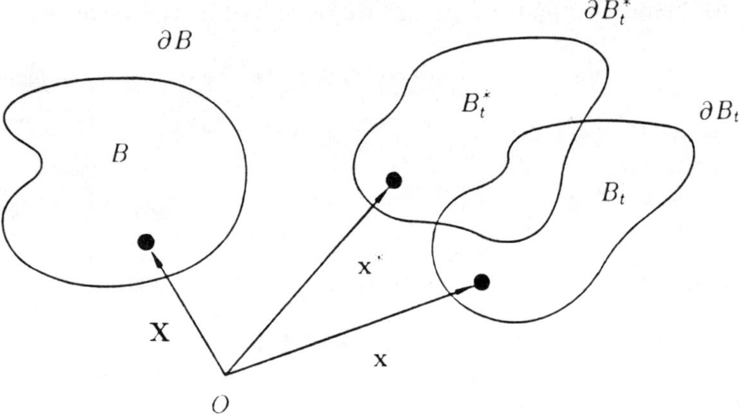

Figure 2.2: The reference configuration, the configuration at time t, and the configuration resulting from the comparison motion at time t.

2.3   The Comparison Motion

In applying Hamilton's principle to a system of particles, the equations of motion were obtained by introducing a comparison motion of the system, Equation (1.15). An identical approach is taken in applying variational methods to a continuous medium.

A motion (2.40) from the reference configuration at time t_1 to a specified configuration at time t_2 will be called *admissible* if it is C^2 on $\overline{B} \times [t_1, t_2]$ and it satisfies prescribed boundary conditions on ∂B. An admissible *comparison motion* will be defined by

$$\mathbf{x}^* = \boldsymbol{\chi}(\mathbf{X}, t) + \varepsilon\boldsymbol{\eta}(\mathbf{X}, t). \qquad (2.71)$$

Here ε is a parameter and $\boldsymbol{\eta}(\mathbf{X}, t)$ is an arbitrary C^2 vector field on $\overline{B} \times [t_1, t_2]$ subject to the requirements that $\boldsymbol{\eta}(\mathbf{X}, t_1) = \mathbf{o}$ and $\boldsymbol{\eta}(\mathbf{X}, t_2) = \mathbf{o}$. The vector field $\boldsymbol{\eta}$ is also subject to the requirement that the comparison motion must satisfy the prescribed boundary conditions on ∂B.

The comparison motion maps the material from B onto a volume B_t^* with surface ∂B_t^* at time t (see Figure 2.2). From Equation (2.71), the velocity, deformation gradient and Jacobian of the comparison motion are

$$\mathbf{v}^* = \mathbf{v} + \varepsilon\dot{\boldsymbol{\eta}},$$

$$\mathbf{F}^* = \mathbf{F} + \varepsilon\frac{\partial\hat{\boldsymbol{\eta}}}{\partial\mathbf{X}}, \quad F_{km}^* = F_{km} + \varepsilon\frac{\partial\hat{\eta}_k}{\partial X_m}, \tag{2.72}$$

$$J^* = \det\mathbf{F}^*.$$

The derivative of J^* with respect to ε is

$$\frac{\partial J^*}{\partial\varepsilon} = \frac{\partial(\det\mathbf{F}^*)}{\partial F_{km}^*}\frac{\partial F_{km}^*}{\partial\varepsilon} = \frac{\partial(\det\mathbf{F}^*)}{\partial F_{km}^*}\frac{\partial\hat{\eta}_k}{\partial X_m}. \tag{2.73}$$

Recalling the notation

$$\delta(\cdot) \equiv \left[\frac{\partial}{\partial\varepsilon}(\cdot)^*\right]_{\varepsilon=0} \tag{2.74}$$

that was introduced in Subsection 1.2.1, (2.26) and (2.73) can be used to obtain the result (see *e.g.* [28])

$$\delta J = \left(\frac{\partial J^*}{\partial\varepsilon}\right)_{\varepsilon=0} = J\,\mathrm{div}\,\boldsymbol{\eta}. \tag{2.75}$$

Therefore, J^* can be written

$$J^* = J(1 + \varepsilon\,\mathrm{div}\,\boldsymbol{\eta}) + O(\varepsilon^2). \tag{2.76}$$

The notation $O(\varepsilon^2)$ means that $[O(\varepsilon^2)/\varepsilon] \to 0$ as $\varepsilon \to 0$.

The density of the comparison motion can be determined from the equation of conservation of mass (2.54).

$$\rho^* = \rho_R/J^*. \tag{2.77}$$

Substituting (2.76) into this equation results in the expression

$$\rho^* = \rho(1 - \varepsilon\,\mathrm{div}\,\boldsymbol{\eta}) + O(\varepsilon^2). \tag{2.78}$$

In applications of Hamilton's principle to a continuous medium, it is often convenient to introduce a *comparison field* for the density in the form

$$\rho^* = \rho(\mathbf{X},t) + \varepsilon r(\mathbf{X},t). \tag{2.79}$$

The preceding two equations show that, as a consequence of the equation of conservation of mass, the scalar field $r = -\rho\,\mathrm{div}\,\boldsymbol{\eta} + O(\varepsilon)$. However, the fields $\boldsymbol{\eta}$ and r can be regarded as *independent* if the equation of conservation of mass is introduced into Hamilton's principle as a constraint (see Subsection 1.2.2). In such cases, it will be assumed that $r(\mathbf{X}, t)$ is an arbitrary C^1 scalar field on $\overline{B} \times [t_1, t_2]$ such that $r(\mathbf{X}, t_1) = 0$ and $r(\mathbf{X}, t_2) = 0$.

As a result of the comparison motion, a volume element dV of B will be mapped onto a volume element dV_t^* at time t. Similarly, a surface element dS of ∂B will be mapped onto a surface element dS_t^* at time t. The relations between these volume and surface elements are obtained from equations (2.52) and (2.59).

$$dV_t^* = J^* dV,$$

$$\mathbf{n}^* dS_t^* = J^* \left(\mathbf{F}^*\right)^{-t} \mathbf{N} dS.$$

(2.80)

Here $\mathbf{n}^*$ is the outward directed unit vector that is normal to ∂B_t^*.

As described in Chapter 1, Hamilton's principle is a postulate concerning the mechanical behavior of a system. The equations of motion are derived from the postulate as necessary conditions. Two examples of the types of analysis involved in obtaining equations of motion from statements of Hamilton's principle for a continuous medium will be presented in the remainder of this section. The methods used are quite similar to those that were used in the case of a system of particles.

In analogy with the kinetic energy of a particle, the kinetic energy of an element dV_t of B_t is $\frac{1}{2}\rho \mathbf{v} \cdot \mathbf{v} dV_t$. Therefore the total kinetic energy of the material occupying the volume B_t is

$$T = \int_{B_t} \tfrac{1}{2}\rho \mathbf{v} \cdot \mathbf{v} dV_t = \int_{B} \tfrac{1}{2}\rho_R \mathbf{v} \cdot \mathbf{v} dV.$$

(2.81)

Consider the integral of T with respect to time from t_1 to t_2,

$$I = \int_{t_1}^{t_2} T \, dt = \int_{t_1}^{t_2} \int_{B} \tfrac{1}{2}\rho_R \mathbf{v} \cdot \mathbf{v} dV \, dt.$$

(2.82)

When it is expressed in terms of the comparison motion (2.71), this integral becomes

$$I^*(\varepsilon) = \int_{t_1}^{t_2} \int_{B} \tfrac{1}{2}\rho_R \mathbf{v}^* \cdot \mathbf{v}^* dV \, dt.$$

(2.83)

Upon taking the derivative with respect to ε,

$$\frac{dI^*(\varepsilon)}{d\varepsilon} = \int_{t_1}^{t_2} \int_{B} \rho_R \mathbf{v}^* \cdot \dot{\boldsymbol{\eta}} dV \, dt,$$

(2.84)

and setting $\varepsilon = 0$, one obtains

$$\left[\frac{dI^*(\varepsilon)}{d\varepsilon}\right]_{\varepsilon = 0} = \int_{t_1}^{t_2} \int_{B} \rho_R \mathbf{v} \cdot \dot{\boldsymbol{\eta}} dV \, dt.$$

(2.85)

Integrating this expression by parts with respect to time gives

$$\int_{t_1}^{t_2} \rho_R \mathbf{v} \cdot \dot{\boldsymbol{\eta}}\, dt = \left[\rho_R \mathbf{v} \cdot \boldsymbol{\eta}\right]_{t_2}^{t_2} - \int_{t_1}^{t_2} \rho_R \mathbf{a} \cdot \boldsymbol{\eta}\, dt. \tag{2.86}$$

Using this result and recalling that $\boldsymbol{\eta}$ vanishes at t_1 and t_2, Equation 2.85 becomes

$$\left[\frac{dI^*(\varepsilon)}{d\varepsilon}\right]_{\varepsilon = 0} = -\int_{t_1}^{t_2}\int_B \rho_R \mathbf{a} \cdot \boldsymbol{\eta}\, dV\, dt = -\int_{t_1}^{t_2}\int_{B_t} \rho \mathbf{a} \cdot \boldsymbol{\eta}\, dV_t\, dt, \tag{2.87}$$

or

$$\delta T = -\int_B \rho_R \mathbf{a} \cdot \delta \mathbf{x}\, dV = -\int_{B_t} \rho \mathbf{a} \cdot \delta \mathbf{x}\, dV_t, \tag{2.88}$$

where

$$\delta \mathbf{x} = \boldsymbol{\eta}. \tag{2.89}$$

Note from Equation (2.71) that this definition of $\delta \mathbf{x}$ is consistent with the notation (2.74).

As a second example, consider the integral

$$C = \int_B \pi \left(J - \frac{\rho_R}{\rho}\right) dV = \int_{B_t} \pi \left(1 - \frac{\rho_R}{\rho J}\right) dV_t. \tag{2.90}$$

This expression is the form in which the equation of conservation of mass will be introduced as a constraint on Hamilton's principle for a continuous medium. The scalar field $\pi(\mathbf{X}, t)$, which is assumed to be C^1 on $\overline{B} \times [t_1, t_2]$, is a Lagrange multiplier. Expressed in terms of the comparison motion (2.71) and the comparison field (2.79), the integral becomes

$$C^*(\varepsilon) = \int_B \pi \left(J^* - \frac{\rho_R}{\rho^*}\right) dV. \tag{2.91}$$

Upon taking the derivative of this equation with respect to ε, setting $\varepsilon = 0$, and using Equation (2.75) to evaluate the derivative of the Jacobian, one obtains the result

$$\delta C = \int_B \pi J \left(\operatorname{div} \boldsymbol{\eta} + \frac{r}{\rho}\right) dV. \tag{2.92}$$

By using the divergence theorem, this result can be written

$$\delta C = \int_{\partial B_t} \pi \mathbf{n} \cdot \delta \mathbf{x}\, dS_t + \int_{B_t} \left(-\operatorname{grad} \pi \cdot \delta \mathbf{x} + \frac{\pi}{\rho}\delta \rho\right) dV_t, \tag{2.93}$$

where $\delta \rho = r$.

2.4   The Fundamental Lemmas

The fundamental lemma of the calculus of variations (see Section 1.1) is the result that is required to obtain differential equations which apply *locally* from a *global* variational statement. In this section, extensions of the fundamental lemma are presented which are appropriate for applications of Hamilton's principle to a continuous medium (see Gurtin [35], pp. 20, 244).

Lemma 1 *Let $\mathcal{W}$ be an inner product space, and consider a C^0 field $\mathbf{f} : \overline{B} \times [t_1, t_2] \to \mathcal{W}$. If the equation*

$$\int_{t_1}^{t_2} \int_B \mathbf{f} \cdot \mathbf{w}\, dV\, dt = 0 \tag{2.94}$$

holds for every C^∞ field $\mathbf{w} : \overline{B} \times [t_1, t_2] \to \mathcal{W}$ that vanishes at time t_1, at time t_2, and on ∂B, then $\mathbf{f} = \mathbf{o}$ on $\overline{B} \times [t_1, t_2]$.

Lemma 2 *Suppose that ∂B consists of complementary regular subsurfaces ∂B_1 and ∂B_2. Let $\mathcal{W}$ be an inner product space, and consider a function $\mathbf{f} : \partial B_2 \times [t_1, t_2] \to \mathcal{W}$ that is piecewise regular and continuous in time. If the equation*

$$\int_{t_1}^{t_2} \int_{\partial B_2} \mathbf{f} \cdot \mathbf{w}\, dS\, dt = 0 \tag{2.95}$$

holds for every C^∞ field $\mathbf{w} : \overline{B} \times [t_1, t_2] \to \mathcal{W}$ that vanishes at time t_1, at time t_2, and on ∂B_1, then $\mathbf{f} = \mathbf{o}$ on $\partial B_2 \times [t_1, t_2]$.

These lemmas are important not only because they are used in obtaining the local forms of the equations of motion from Hamilton's principle, but also because they impose smoothness requirements on the fields describing the material. Both of these aspects were illustrated in the case of a system of particles in Chapter 1.

Due to their importance, a proof of Lemma 1 given by Gurtin ([35], p. 224) will be presented. The proof proceeds by assuming that the field $\mathbf{f}$ does not vanish at some point $\mathbf{X}_0, t_0$ in $B \times (t_1, t_2)$, and then constructing a suitable field $\mathbf{w}$ such that Equation (2.94) is violated.

Let $\{\mathbf{e}_k\}$ be an orthonormal basis for $\mathcal{W}$, so that $\mathbf{f}$ can be written $\mathbf{f} = f_k \mathbf{e}_k$. Assume that $f_k(\mathbf{X}_0, t_0) > 0$ for some value of k and some point $\mathbf{X}_0, t_0$ in $B \times (t_1, t_2)$. Let α be a positive scalar. Denote the open interval of time $(t_0 - \alpha, t_0 + \alpha)$ by T_α, and denote the open region $|\mathbf{X}_0 - \mathbf{X}| < \alpha$ by Ω_α. Because of the continuity of $\mathbf{f}$, there is an α such that $f_k(\mathbf{X}, t) > 0$ in $\Omega_\alpha \times T_\alpha$. Define $\beta(t)$ to be a scalar function on $[t_1, t_2]$ which is C^∞ and has the property that $\beta(t) > 0$ if t is in T_α and $\beta(t) = 0$ otherwise. Define $\gamma(\mathbf{X})$ to be a scalar field on $\overline{B}$ which is C^∞ and has the property that $\gamma(\mathbf{X}) > 0$ if $\mathbf{X}$ is in Ω_α and $\gamma(\mathbf{X}) = 0$ otherwise.[7] Then define

$$\mathbf{w} = \beta(t)\gamma(\mathbf{X})\mathbf{e}_k. \tag{2.96}$$

The vector field $\mathbf{w}$ is C^∞ on $\overline{B} \times [t_1, t_2]$ and vanishes at time t_1, at time t_2, and on ∂B. It has been constructed so that

$$\int_{t_1}^{t_2} \int_B \mathbf{f} \cdot \mathbf{w}\, dV\, dt = \int_{T_\alpha} \int_{\Omega_\alpha} f_k \beta(t)\gamma(\mathbf{X})\, dV\, dt > 0, \tag{2.97}$$

which violates (2.94). Therefore $\mathbf{f}$ must vanish in $B \times (t_1, t_2)$. Since the field $\mathbf{f}$ is continuous on $\overline{B} \times [t_1, t_2]$, it must vanish on $\overline{B} \times [t_1, t_2]$.

Minor variations of Lemmas 1 and 2 will also be used.

Lemma 3 *Let the motion (2.40) be C^2 on $\overline{B} \times [t_1, t_2]$. Let $\mathcal{W}$ be an IPS, and consider a C^0 field $\mathbf{f} : \overline{B} \times [t_1, t_2] \to \mathcal{W}$. If the equation*

$$\int_{t_1}^{t_2} \int_{B_t} \mathbf{f} \cdot \mathbf{w}\, dV_t dt = 0 \tag{2.98}$$

holds for every C^∞ field $\mathbf{w} : \overline{B} \times [t_1, t_2] \to \mathcal{W}$ that vanishes at time t_1, at time t_2, and on ∂B, then $\mathbf{f}(\mathbf{x}, t) = \mathbf{o}$ on $\overline{B}_t \times [t_1, t_2]$.

To prove this result, (2.98) can be written

$$\int_{t_1}^{t_2} \int_B J\mathbf{f} \cdot \mathbf{w}\, dV\, dt = 0. \tag{2.99}$$

Since $J\mathbf{f}$ is continuous on $\overline{B} \times [t_1, t_2]$, Lemma 1 requires that $J\mathbf{f} = \mathbf{o}$ on $\overline{B} \times [t_1, t_2]$. Since $J > 0$, $\mathbf{f}(\mathbf{X}, t) = \mathbf{o}$ on $\overline{B} \times [t_1, t_2]$, so $\mathbf{f}(\mathbf{x}, t) = \mathbf{o}$ on $\overline{B}_t \times [t_1, t_2]$.

Lemma 4 *The complementary regular subsurfaces ∂B_1 and ∂B_2 will be mapped onto surfaces ∂B_{t1} and ∂B_{t2} by the motion (2.40). Let the motion (2.40) be C^2 on $\overline{B} \times [t_1, t_2]$. Let $\phi(\mathbf{X}, t)$ be a scalar function defined on $\partial B_2 \times [t_1, t_2]$ that is piecewise regular and continuous in time. If the equation*

$$\int_{t_1}^{t_2} \int_{\partial B_{t2}} \phi\mathbf{n} \cdot \mathbf{w}\, dS_t dt = 0 \tag{2.100}$$

holds for every vector field $\mathbf{w}$ that is C^∞ on $\overline{B} \times [t_1, t_2]$ and vanishes at time t_1, at time t_2, and on ∂B_1, then $\phi = 0$ on $\partial B_{t2} \times [t_1, t_2]$.

---

[7]The existence of functions $\beta(t)$ and $\gamma(\mathbf{X})$ having these properties can be demonstrated (Gurtin [35], p. 19).

The proof is similar to that of Lemma 3. Equation (2.100) can be written

$$\int_{t_1}^{t_2} \int_{\partial B_2} \phi J \mathbf{F}^{-t} \mathbf{N} \cdot \mathbf{w} \, dS \, dt = 0. \tag{2.101}$$

Since $\phi J \mathbf{F}^{-t}$ is continuous on $\overline{B} \times [t_1, t_2]$, $\phi J \mathbf{F}^{-t} \mathbf{N}$ is piecewise regular and continuous in time on $\partial B_2 \times [t_1, t_2]$. Therefore, Lemma 2 requires that $\phi J \mathbf{F}^{-t} \mathbf{N} = \mathbf{o}$ on $\partial B_2 \times [t_1, t_2]$. From (2.59), this implies that $\phi = 0$ on $\partial B_{t2} \times [t_1, t_2]$.

3 Mechanics of continuous media

Applications of Hamilton's principle to deformable continuous media are discussed in this chapter. Statements of the principle for continuous media are formally very similar to those for systems of particles, and certainly were motivated by them. However, it should be emphasized that the statements for continuous media stand as independent postulates; they are not *derived* from Hamilton's principle for a system of particles. It will be shown that the local forms of the equations of motion for continuous media and their associated boundary conditions are obtained as necessary conditions implied by Hamilton's principle. The classical theories of fluid and solid mechanics will be described, and also two recent theories of materials with microstructure. It will be shown that postulates of Hamilton's principle that have been introduced to obtain more general theories are natural and well motivated extensions of the classical theories.

Problems in the mechanics of continuous media usually involve nonconservative forces, and it is frequently convenient to include constraints in statements of Hamilton's principle. Therefore, the postulates that will be introduced in this work will be expressed in the same form as the second form of Hamilton's principle for a system of particles stated on page 12. They will be developed by the heuristic approach of identifying terms associated with the mechanics of continuous media which are analogous to the terms which appear in Equation (1.48). In one example which does not involve nonconservative forces, Hamilton's principle will also be expressed in the first form stated on page 6.

3.1 The Classical Theories

In this section, theories are discussed in which the mechanical behavior of a material is completely described by its motion

$$\mathbf{x} = \chi(\mathbf{X}, t). \tag{3.1}$$

Hamilton's principle will be postulated for a finite amount of material which occupies a bounded regular region B in a prescribed reference configuration at time t_1. As the material undergoes a motion (3.1), it will occupy a volume B_t at each time t. Therefore B_t is called a *material volume;* it contains the same material at each time t.

Throughout this section, an *admissible motion* will refer to a motion (3.1) of the material from the prescribed reference configuration at time t_1 to a prescribed configuration at time t_2 which is C^2 on $\overline{B} \times [t_1, t_2]$ and which satisfies prescribed boundary conditions on ∂B. A *comparison motion* will refer to an admissible motion

$$\mathbf{x}^* = \boldsymbol{\chi}(\mathbf{X}, t) + \varepsilon\boldsymbol{\eta}(\mathbf{X}, t), \tag{3.2}$$

where $\boldsymbol{\eta}(\mathbf{X}, t)$ is an arbitrary C^2 vector field on $\overline{B} \times [t_1, t_2]$ subject to the requirements that $\boldsymbol{\eta}(\mathbf{X}, t_1) = \mathbf{o}$ and $\boldsymbol{\eta}(\mathbf{X}, t_2) = \mathbf{o}$.

3.1.1 Ideal Fluids

The terms *ideal* or *inviscid* fluid refer to a model of fluid behavior in which the effects of viscosity are neglected. This is the simplest model for a continuous medium. Two cases, compressible and incompressible fluids, will be treated.

Consider how one might postulate Hamilton's principle for an ideal fluid in a form analogous to the first form for a system of particles stated on page 12. An admissible motion and comparison motion of the fluid are given by Equations (3.1) and (3.2). It will be assumed that there is no geometrical constraint on the motion of the fluid on ∂B.

The density $\rho(\mathbf{X}, t)$ of the fluid will be assumed to be a C^1 scalar field on $\overline{B} \times [t_1, t_2]$. In Section 2.3 an admissible *comparison density field* was defined by

$$\rho^* = \rho(\mathbf{X}, t) + \varepsilon r(\mathbf{X}, t), \tag{3.3}$$

where $r(\mathbf{X}, t)$ is an arbitrary C^1 scalar field on $\overline{B} \times [t_1, t_2]$ subject to the requirements that $r(\mathbf{X}, t_1) = 0$ and $r(\mathbf{X}, t_2) = 0$.

Consider the individual terms in Equation (1.48).

The potential energy U. If the fluid is compressible, it can store potential energy in the form of energy of deformation. The deformation of a fluid is expressed in terms of its change in density from a reference state. Therefore, let it be assumed that there is a scalar function $e(\rho)$, the *internal energy*, which is defined such that the potential energy of each element dV_t of the fluid contained in B_t is $\rho e(\rho) dV_t$. It will be assumed that the second derivative of the function $e(\rho)$ exists and is continuous. In modern terminology, the assumption that the internal energy depends only on the density of the fluid is a *constitutive assumption* that characterizes an *elastic fluid*. Therefore the total potential energy of the fluid contained in B_t is

$$U = \int_{B_t} \rho e(\rho)\, dV_t. \tag{3.4}$$

The virtual work term δW. The external forces acting on the fluid will be introduced by means of virtual work terms. Let there be a prescribed vector field $\mathbf{b}(\mathbf{X},t)$, the *body force*, that is C^0 on $\overline{B} \times [t_1,t_2]$ and defined such that the external force exerted on a volume element dV_t of the fluid contained in B_t is $\rho \mathbf{b}\, dV_t$. This field represents external forces which are distributed over the volume of the fluid, such as its weight. The virtual work done by this force will be expressed in the form $\rho \mathbf{b}\, dV_t \cdot \delta \mathbf{x}$, which is clearly analogous to Equation (1.42). The virtual work on the fluid contained in B_t is

$$\int_{B_t} \rho \mathbf{b} \cdot \delta \mathbf{x}\, dV_t.$$

It will also be assumed that there is a prescribed scalar field $p_0(\mathbf{X},t)$, the *external pressure*, that is continuous in time and piecewise regular on $\partial B \times [t_1,t_2]$ and defined such that the external force exerted on an area element dS_t of ∂B_t is $-p_0 \mathbf{n}\, dS_t$. The virtual work will be written $-p_0 \mathbf{n}\, dS_t \cdot \delta \mathbf{x}$, and the virtual work on the fluid contained in B_t is

$$-\int_{\partial B_t} p_0 \mathbf{n} \cdot \delta \mathbf{x}\, dS_t.$$

Therefore, the virtual work done on the fluid by external forces is postulated to be of the form

$$\delta W = \int_{B_t} \rho \mathbf{b} \cdot \delta \mathbf{x}\, dV_t - \int_{\partial B_t} p_0 \mathbf{n} \cdot \delta \mathbf{x}\, dS_t. \tag{3.5}$$

The constraint C. The motion of the fluid and its density field are related through the equation of conservation of mass. The comparison motion (3.2) and the comparison density field (3.3) can be regarded as independent if the equation of conservation of mass (2.54) is introduced into Hamilton's principle as a constraint. The constraint will be written in the form

$$C = \int_{B_t} \pi \left(1 - \frac{\rho_R}{\rho J} \right) dV_t, \tag{3.6}$$

where the unknown field $\pi(\mathbf{X},t)$, which is assumed to be C^1 on $\overline{B} \times [t_1,t_2]$, is a Lagrange multiplier.

The kinetic energy T. The kinetic energy of the fluid is

$$T = \int_{B_t} \tfrac{1}{2} \rho \mathbf{v} \cdot \mathbf{v}\, dV_t. \tag{3.7}$$

Hamilton's principle for an elastic ideal fluid states: *Among comparison motions* (3.2) *and comparison density fields* (3.3), *the actual fields are such that*

$$\int_{t_1}^{t_2} [\delta(T - U) + \delta W + \delta C]dt = 0. \tag{3.8}$$

It was shown in Section 2.3 [Equations (2.88) and (2.93)] that the terms δT and δC can be written

$$\delta T = -\int_{B_t} \rho \mathbf{a} \cdot \delta \mathbf{x} dV_t, \tag{3.9}$$

$$\delta C = \int_{\partial B_t} \pi \mathbf{n} \cdot \delta \mathbf{x} dS_t + \int_{B_t} \left(-\operatorname{grad} \pi \cdot \delta \mathbf{x} + \frac{\pi}{\rho}\delta\rho\right) dV_t. \tag{3.10}$$

The potential energy is

$$U = \int_{B_t} \rho e(\rho) dV_t = \int_B \rho_R e(\rho) dV. \tag{3.11}$$

In terms of the comparison density field (3.3), this expression is

$$U^* = \int_B \rho_R e^* dV, \tag{3.12}$$

where $e^* = e(\rho^*)$. The derivative with respect to ε is

$$\frac{dU^*}{d\varepsilon} = \int_B \rho_R \frac{de^*}{d\rho^*}\frac{\partial\rho^*}{\partial\varepsilon}dV = \int_B \rho_R \frac{de^*}{d\rho^*}r dV, \tag{3.13}$$

so that the variation of the potential energy is

$$\delta U = \int_B \rho_R \frac{de}{d\rho}\delta\rho dV = \int_{B_t} \rho \frac{de}{d\rho}\delta\rho dV_t. \tag{3.14}$$

Upon substituting the expressions (3.5), (3.9), (3.10), and (3.14) into Equation (3.8), it can be written

$$\int_{t_1}^{t_2} \left[\int_{B_t} (-\rho\mathbf{a} - \operatorname{grad}\pi + \rho\mathbf{b}) \cdot \delta\mathbf{x} dV_t \right.$$

$$+ \int_{B_t} \left(\frac{\pi}{\rho} - \rho\frac{de}{d\rho}\right)\delta\rho dV_t \tag{3.15}$$

$$\left. + \int_{\partial B_t} (\pi - p_0)\mathbf{n} \cdot \delta\mathbf{x} dS_t \right] dt = 0.$$

The equation of motion and boundary condition for the fluid are deduced from this equation by applying Lemmas 3 and 4 of Section 2.4. Since the fields $\boldsymbol{\eta} = \delta\mathbf{x}$ and $r = \delta\rho$ are arbitrary, it can be assumed that $\delta\rho = 0$ on $\overline{B} \times [t_1, t_2]$ and that $\delta\mathbf{x} = 0$

on $\partial B \times [t_1, t_2]$. As a result, the second and third integrals in (3.15) vanish. Then applying Lemma 3 to the remaining term yields the differential equation

$$\rho \mathbf{a} = - \operatorname{grad} \pi + \rho \mathbf{b} \quad \text{on } \overline{B}_t \times [t_1, t_2]. \tag{3.16}$$

This is called the equation of *balance of linear momentum*. Next, assuming that $\delta \mathbf{x} = 0$ on $\overline{B} \times [t_1, t_2]$ and applying Lemma 3 yields the equation

$$\pi = \rho^2 \frac{de}{d\rho} \quad \text{on } \overline{B}_t \times [t_1, t_2]. \tag{3.17}$$

This equation determines the Lagrange multiplier π as a function of the density of the fluid. Finally, applying Lemma 4 provides the boundary condition

$$\pi = p_0 \quad \text{on } \partial B_t \times [t_1, t_2]. \tag{3.18}$$

A physical interpretation of the Lagrange multiplier π can be gained by writing Equation (3.15) in terms of a volume of fluid B_t' that is contained within B_t (Figure 3.1).

$$\int_{t_1}^{t_2} \left[\int_{B_t'} (-\rho \mathbf{a} - \operatorname{grad} \pi + \rho \mathbf{b}) \cdot \delta \mathbf{x} dV_t \right.$$
$$+ \int_{B_t'} \left(\frac{\pi}{\rho} - \rho \frac{de}{d\rho} \right) \delta \rho dV_t \tag{3.19}$$
$$+ \left. \int_{\partial B_t'} (\pi - p) \mathbf{n} \cdot \delta \mathbf{x} dS_t \right] dt = 0.$$

Here the term $-p\mathbf{n}$ is the normal traction exerted on the fluid within B_t' by the fluid exterior to B_t'; that is, p is the pressure of the fluid. The function p is not prescribed, but is a *constitutive function* which is assumed to be C^1 on $\overline{B} \times [t_1, t_2]$. By the same procedure that was applied to (3.15), (3.19) implies Equations (3.16) and (3.17) on $\overline{B}_t' \times [t_1, t_2]$, and the equation

$$\pi = p \quad \text{on } \partial B_t' \times [t_1, t_2]. \tag{3.20}$$

Thus *the Lagrange multiplier π is the pressure of the fluid.* Furthermore, observe from Equation (3.17) that *Hamilton's principle yields the constitutive equation for the pressure of the fluid in terms of the internal energy.*

At this point, the usual method of determining the equation of motion for an ideal fluid will be sketched for the purpose of comparison with Hamilton's principle. The approach used is to write a postulate for an arbitrary material volume of the

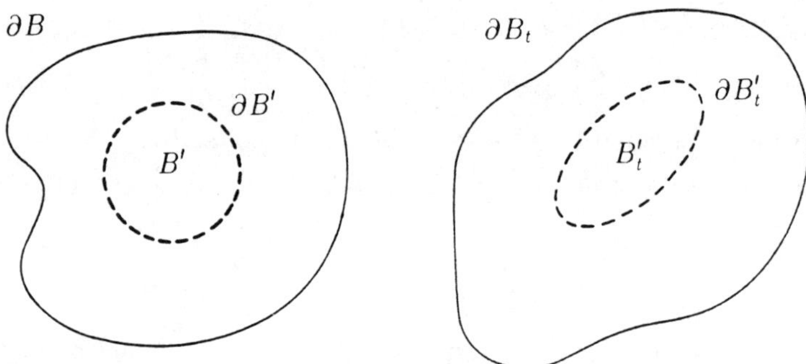

Figure 3.1: A volume of fluid B'_t that is contained within B_t.

fluid that is analogous to Newton's second law for a system of particles (see *e.g.* Gurtin [36], pp. 105–110). Let B'_t be an arbitrary volume contained within B_t (Figure 3.1). The *linear momentum* of an element dV_t of B'_t is the product of its mass and velocity, $\rho\mathbf{v}dV_t$. It is postulated that, at an arbitrary instant in time, the rate of change of the total linear momentum of the fluid contained within B'_t is equal to the total external force exerted on the fluid.

$$\frac{d}{dt}\int_{B'_t}\rho\mathbf{v}dV_t = \int_{B'_t}\rho\mathbf{b}dV_t - \int_{\partial B'_t}p\mathbf{n}dS_t. \tag{3.21}$$

As a consequence of the equation of conservation of mass and Reynolds' transport theorem (see *e.g.* Gurtin [36], pp. 78–79),

$$\frac{d}{dt}\int_{B'_t}\rho\mathbf{v}dV_t = \int_{B'_t}\rho\mathbf{a}dV_t. \tag{3.22}$$

Using this result and the divergence theorem, Equation (3.21) can be written

$$\int_{B'_t}(\rho\mathbf{a} + \operatorname{grad} p - \rho\mathbf{b})dV_t = 0. \tag{3.23}$$

Since the volume B'_t is arbitrary, this equation implies that

$$\rho\mathbf{a} = -\operatorname{grad} p + \rho\mathbf{b} \quad \text{on } \overline{B}_t \times [t_1, t_2]. \tag{3.24}$$

It is clear from this derivation why this equation is referred to as the balance of linear momentum.

This *direct* method of obtaining this equation is simpler than Hamilton's principle, although if the derivation of Reynolds' transport theorem is regarded as an integral part of the process, the difference is not so pronounced. Nevertheless, their relative complexity is a criticism that has been made of variational methods in continuum mechanics. Note, however, that the direct method does not yield Equation (3.17). Other advantages of Hamilton's principle, particularly in connection with its ability to incorporate constraints, will be illustrated in subsequent examples. The author regards direct and variational methods as complementary, not competitive. In some cases one method is more advantageous and in some the other, and often both methods lend insight to a given problem.

If the constitutive relation for the internal energy as a function of the density is known, Equations (2.58), (3.16) and (3.17) provide a system of equations to determine the density field ρ, the velocity field $\mathbf{v}$, and the pressure field $\pi = p$. Although this is a simple theory, it is the one used in the study of aerodynamics to analyze high speed flows, except in regions (such as boundary layers and wakes) where the effects of viscosity cannot be neglected. In this application, Equation (3.17) is usually assumed to be the isentropic relation

$$\frac{p}{\rho^\gamma} = \text{constant}, \tag{3.25}$$

where γ, the *ratio of specific heats*, is treated as a constant. When it is linearized in terms of small perturbations, this theory is also used in the study of the propagation of acoustic waves. Let

$$\mathbf{v} = \tilde{\mathbf{v}},$$
$$\rho = \rho_R + \tilde{\rho}, \tag{3.26}$$

where $\tilde{\mathbf{v}}$ and $\tilde{\rho}$ are small perturbations and ρ_R is assumed to be homogeneous. Using (3.25), Equations (2.58) and (3.16) can be written (in the absense of the body force)

$$\frac{\partial \tilde{\rho}}{\partial t} + \rho_R \operatorname{div} \tilde{\mathbf{v}} = 0,$$
$$\rho_R \frac{\partial \tilde{\mathbf{v}}}{\partial t} = -(\gamma p_R / \rho_R) \operatorname{grad} \tilde{\rho}. \tag{3.27}$$

Eliminating $\tilde{\mathbf{v}}$ from these two equations yields the linear wave equation

$$\frac{\partial^2 \tilde{\rho}}{\partial t^2} = \alpha^2 \nabla^2 \tilde{\rho}, \tag{3.28}$$

where $\alpha = (\gamma p_R / \rho_R)^{\frac{1}{2}}$ is the sound speed.

If the fluid is incompressible, $\rho = \rho_R = $ constant. In this case there is no energy of deformation, so the potential energy $U = 0$. The equation of conservation of mass (2.54) is $J = 1$, so the constraint (3.6) reduces to

$$C = \int_{B_t} \pi \left(1 - \frac{1}{J} \right) dV_t. \tag{3.29}$$

The other terms in Equation (3.8) are unchanged. By using the same procedure as in the case of the compressible fluid, (3.8) leads to the equation of balance of linear momentum

$$\rho \mathbf{a} = -\operatorname{grad} \pi + \rho \mathbf{b} \quad \text{on } \overline{B}_t \times [t_1, t_2] \tag{3.30}$$

and the boundary condition

$$\pi = p_0 \quad \text{on } \partial B_t \times [t_1, t_2]. \tag{3.31}$$

This equation of balance of linear momentum and boundary condition are identical to those obtained in the case of a compressible fluid, but there is no equivalent to Equation (3.17). The pressure π is not a constitutive function of the density. In the case of an incompressible fluid, Equation (2.57) reduces to

$$\operatorname{div} \mathbf{v} = 0. \tag{3.32}$$

Equations (3.30) and (3.32) provide two equations to determine the velocity field $\mathbf{v}$ and the pressure field π.

Applications of Hamilton's principle to ideal fluids are discussed by Eckart [24], Herivel [44], Lanczos [50], Leech [51], Serrin [65], and Taub [67].

3.1.2   Elastic Solids

An elastic solid can be characterized by the assumption that the internal energy is a function of the deformation gradient $\mathbf{F}$, so that the potential energy of the material contained in B_t is

$$U = \int_{B_t} \rho e(\mathbf{F}) dV_t = \int_B \rho_R e(\mathbf{F}) dV. \tag{3.33}$$

It will be assumed that the function e is differentiable on a suitable open domain of its argument and that $\operatorname{DIV}\left(\dfrac{\partial e}{\partial \mathbf{F}} \right)$ is continuous on $\overline{B} \times [t_1, t_2]$.

Suppose that the surface ∂B consists of complementary regular subsurfaces ∂B_1 and ∂B_2, and that the motion of the material is prescribed on ∂B_1. Let there be no constraint on the motion of the material on ∂B_2.

40

Let there be a prescribed vector function $t_0(x, t)$, the *external traction*, that is defined on $\partial B_{t2} \times [t_1, t_2]$ such that the external force exerted on an element dS_t of ∂B_{t2} is $t_0 dS_t$. Then define a vector field $s_0(X, t)$ by $s_0 dS = t_0 dS_t$, and let s_0 be assumed to be continuous in time and piecewise regular on $\partial B_2 \times [t_1, t_2]$. The virtual work is $t_0 dS_t \cdot \delta x = s_0 dS \cdot \delta x$, and the total virtual work done by external forces on the material contained in B_t is

$$\delta W = \int_B \rho_R b \cdot \delta x dV + \int_{\partial B_2} s_0 \cdot \delta x dS, \tag{3.34}$$

where the body force b is defined as in the preceding subsection. Note that δx must vanish on ∂B_1 since the comparison motion (3.2) must satisfy the prescribed boundary conditions on ∂B.

The kinetic energy of the material contained in B_t is

$$T = \int_B \tfrac{1}{2} \rho_R v \cdot v \, dV. \tag{3.35}$$

Hamilton's principle for an elastic material states: *Among admissible comparison motions (3.2), the actual motion of the material is such that*

$$\int_{t_1}^{t_2} [\delta(T - U) + \delta W] \, dt = 0. \tag{3.36}$$

To determine the variation of the internal energy, it is first expressed in terms of the comparison motion,

$$U^* = \int_B \rho_R e\left(F^*\right) dV. \tag{3.37}$$

The derivative of this expression with respect to ε is

$$\frac{\partial U^*}{\partial \varepsilon} = \int_B \rho_R \frac{de^*}{dF^*} \cdot \frac{\partial F^*}{\partial \varepsilon} dV = \int_B \rho_R \frac{de^*}{dF^*} \cdot \frac{\partial \eta}{\partial X} dV, \tag{3.38}$$

where $e^* = e(F^*)$. In terms of components, this equation is

$$\frac{\partial U^*}{\partial \varepsilon} = \int_B \rho_R \frac{\partial e^*}{\partial F_{km}^*} \frac{\partial F_{km}^*}{\partial \varepsilon} dV = \int_B \rho_R \frac{\partial e^*}{\partial F_{km}^*} \frac{\partial \eta_k}{\partial X_m} dV. \tag{3.39}$$

Therefore, the variation is

$$\delta U = \left[\frac{\partial U^*}{\partial \varepsilon}\right]_{\varepsilon=0} = \int_B S \cdot \frac{\partial \eta}{\partial X} dV, \tag{3.40}$$

where

$$S = \rho_R \frac{de}{dF}, \quad S_{km} = \rho_R \frac{\partial e}{\partial F_{km}}. \tag{3.41}$$

The linear transformation $\mathbf{S}$ is called the *first Piola-Kirchhoff stress*. By means of the divergence theorem, (3.40) can be written

$$\delta U = \int_{\partial B_2} \mathbf{S}\mathbf{N} \cdot \delta\mathbf{x}\,dS - \int_B \text{DIV }\mathbf{S} \cdot \delta\mathbf{x}\,dV. \tag{3.42}$$

Substituting this expression, (2.88), and (3.34) into Equation (3.36), it assumes the form

$$\int_{t_1}^{t_2} \left[\int_B (-\rho_R \mathbf{a} + \text{DIV }\mathbf{S} + \rho_R \mathbf{b}) \cdot \delta\mathbf{x}\,dV \right.$$
$$\left. + \int_{\partial B_2} (\mathbf{s}_0 - \mathbf{S}\mathbf{N}) \cdot \delta\mathbf{x}\,dS \right] dt = 0. \tag{3.43}$$

Invoking Lemmas 1 and 2 of Section 2.4, this equation yields the equation of balance of linear momentum

$$\rho_R \mathbf{a} = \text{DIV }\mathbf{S} + \rho_R \mathbf{b} \quad \text{on } \overline{B} \times [t_1, t_2], \tag{3.44}$$

and the boundary condition

$$\mathbf{S}\mathbf{N} = \mathbf{s}_0 \quad \text{on } \partial B_2 \times [t_1, t_2]. \tag{3.45}$$

When the constitutive relation $e(\mathbf{F})$ is specified, (3.41) and (3.44) may be used to determine the displacement field $\mathbf{u}$ and the Piola-Kirchhoff stress $\mathbf{S}$. The constitutive relations for elastic materials are discussed by Gurtin ([35], Chapter C, [36], Chapters IX, X), Truesdell and Noll ([70], Chapters C, D), and Truesdell and Toupin ([71], pp. 723–727). The *linear theory of elasticity* is obtained by assuming that e is a quadratic form in the linear strain $\mathbf{E}$,

$$\rho_R e = \tfrac{1}{2} A_{ijkm} E_{ij} E_{km}, \tag{3.46}$$

where the coefficients A_{ijkm} are constants. If the material is *isotropic*, it can be shown that

$$A_{ijkm} = \lambda \delta_{ij}\delta_{km} + \mu(\delta_{ik}\delta_{jm} + \delta_{im}\delta_{jk}), \tag{3.47}$$

where λ and μ are the *Lamé constants*. In this case the constitutive equation for the Piola-Kirchhoff stress is

$$\mathbf{S} = \lambda(\,\text{tr }\mathbf{E})\mathbf{1} + 2\mu\mathbf{E}. \tag{3.48}$$

By expressing the variation of the potential energy (3.40) as an integral over B_t, it can be written

$$\delta U = \int_{B_t} \mathbf{T} \cdot \frac{\partial \boldsymbol{\eta}}{\partial \mathbf{x}}\,dV_t, \tag{3.49}$$

where the *Cauchy stress* $\mathbf{T}$ is defined by

$$\mathbf{T} = \frac{1}{J}\mathbf{S}\mathbf{F}^t, \quad T_{km} = \frac{1}{J}S_{kj}\frac{\partial \chi_m}{\partial X_j}. \tag{3.50}$$

By the use of the expression (3.49) for δU, Equation (3.43) can be written

$$\int_{t_1}^{t_2}\left[\int_{B_t}(-\rho\mathbf{a} + \operatorname{div}\mathbf{T} + \rho\mathbf{b})\cdot\delta\mathbf{x}dV_t\right.$$

$$\left. + \int_{\partial B_{t2}}(\mathbf{t}_0 - \mathbf{T}\mathbf{n})\cdot\delta\mathbf{x}dS_t\right]dt = 0, \tag{3.51}$$

resulting in the equation of balance of linear momentum

$$\rho\mathbf{a} = \operatorname{div}\mathbf{T} + \rho\mathbf{b} \quad \text{on } \overline{B}_t \times [t_1, t_2], \tag{3.52}$$

and the boundary condition

$$\mathbf{T}\mathbf{n} = \mathbf{t}_0 \quad \text{on } \partial B_{t2} \times [t_1, t_2]. \tag{3.53}$$

There are some elastic materials, of which rubber is the best known example, for which the assumption that the material is incompressible can be a useful approximation. The equations governing an incompressible elastic material can be obtained by introducing the constraint (3.29) into Hamilton's principle. When the variation of (3.29) is included in Equation (3.51), the resulting equation of balance of linear momentum is

$$\rho\mathbf{a} = \operatorname{div}\left(-\pi\mathbf{1} + \mathbf{T}\right) + \rho\mathbf{b} \quad \text{on } \overline{B}_t \times [t_1, t_2], \tag{3.54}$$

and the boundary condition is

$$(-\pi\mathbf{1} + \mathbf{T})\mathbf{n} = \mathbf{t}_0 \quad \text{on } \partial B_{t2}. \tag{3.55}$$

In the case of an incompressible elastic material, there is an additional governing equation, the constraint

$$J = 1, \tag{3.56}$$

and an additional unknown field, the *pressure* π.

If the external forces acting on an elastic material are conservative, Hamilton's principle can be stated in a manner analogous to the first form for a system of particles on page 6 (see *e.g.* Washizu [73]). Suppose that there exist scalar fields $\psi_b(\mathbf{x}, t)$ and $\psi_s(\mathbf{x}, t)$ such that

$$\mathbf{b} = -\operatorname{grad}\psi_b, \quad \mathbf{s}_0 = -\operatorname{grad}\psi_s. \tag{3.57}$$

Then Hamilton's principle for an elastic material can be stated: *Among admissible motions, the actual motion of the material is such that the integral*

$$I = \int_{t_1}^{t_2} (T - U - U_e)\,dt \qquad (3.58)$$

is stationary in comparison with neighboring admissible motions. The term U_e is defined by

$$U_e = \int_B \rho_R \psi_b\,dV + \int_{\partial B} \psi_s\,dS. \qquad (3.59)$$

In terms of the comparison motion, the integral (3.58) is

$$I^*(\varepsilon) = \int_{t_1}^{t_2} (T^* - U^* - U_e^*)\,dt, \qquad (3.60)$$

where

$$U_e^* = \int_B \rho_R \psi_b^*\,dV + \int_{\partial B} \psi_s^*\,dS, \qquad (3.61)$$

$\psi_b^* = \psi_b(\mathbf{x}^*, t)$, and $\psi_s^* = \psi_s(\mathbf{x}^*, t)$. The derivative of this expression with respect to ε is

$$\begin{aligned}
\frac{\partial U_e^*}{\partial \varepsilon} &= \int_B \rho_R \frac{\partial \psi_b^*}{\partial \mathbf{x}^*} \cdot \frac{\partial \mathbf{x}^*}{\partial \varepsilon}\,dV + \int_{\partial B} \frac{\partial \psi_s^*}{\partial \mathbf{x}^*} \cdot \frac{\partial \mathbf{x}^*}{\partial \varepsilon}\,dS \\
&= \int_B \rho_R \frac{\partial \psi_b^*}{\partial \mathbf{x}^*} \cdot \boldsymbol{\eta}\,dV + \int_{\partial B_2} \frac{\partial \psi_s^*}{\partial \mathbf{x}^*} \cdot \boldsymbol{\eta}\,dS,
\end{aligned} \qquad (3.62)$$

and, using (3.57), the value of this derivative when $\varepsilon = 0$ is

$$\left[\frac{\partial U_e^*}{\partial \varepsilon} \right]_{\varepsilon=0} = -\int_B \rho_R \mathbf{b} \cdot \boldsymbol{\eta}\,dV - \int_{\partial B_2} \mathbf{s}_0 \cdot \boldsymbol{\eta}\,dS. \qquad (3.63)$$

Therefore the first form of Hamilton's principle for an elastic material implies that

$$\begin{aligned}
\left[\frac{dI^*(\varepsilon)}{d\varepsilon} \right]_{\varepsilon=0} = \int_{t_1}^{t_2} \Bigg[\; &-\int_B \rho_R \mathbf{a} \cdot \boldsymbol{\eta}\,dV + \int_B \operatorname{DIV} S \cdot \boldsymbol{\eta}\,dV \\
&- \int_{\partial B_2} \mathbf{SN} \cdot \boldsymbol{\eta}\,dS + \int_B \rho_R \mathbf{b} \cdot \boldsymbol{\eta}\,dV \\
&+ \int_{\partial B_2} \mathbf{s}_0 \cdot \boldsymbol{\eta}\,dS \Bigg] dt = 0,
\end{aligned} \qquad (3.64)$$

which is identical to Equation (3.43).

The application of Hamilton's principle to elastic materials is discussed by Gurtin ([3], pp. 223–226), Love ([53], Chapter VII), Washizu [73], and Weinstock [74].

44

3.1.3 Inelastic Materials

The theories of elastic materials discussed in the preceding two subsections are very special due to the assumptions that are made concerning the functional form of the internal energy. Those assumptions restrict the application of the resulting equations, *a priori*, to elastic materials. For dissipative media, such as viscous fluids, viscoelastic materials, or thermoelastic materials, a more general approach is necessary.

Hamilton's principle can be stated for an *arbitrary* continuous medium, restricted only by the assumption that it does not exhibit microstructural effects. In place of the variation of the internal energy, a virtual work term of the form

$$- \int_B \mathbf{S} \cdot \delta \mathbf{F} \, dV \tag{3.65}$$

is introduced, where the linear transformation $\mathbf{S}$ is a *constitutive variable* subject only to the requirement that $\mathbf{S}$ and $\mathrm{DIV}\,\mathbf{S}$ be continuous on $\overline{B} \times [t_1, t_2]$. *No assumption is made concerning the dependence of $\mathbf{S}$ upon the motion or deformation of the material.* It is only assumed that work is done when the deformation gradient of the material changes, and $\mathbf{S}$ is the associated generalized force. An understanding of this point is essential to an appreciation of the applicability of Hamilton's principle to continuum mechanics.

Let the virtual work on the material contained in B_t be written

$$\delta W = - \int_B \mathbf{S} \cdot \delta \mathbf{F} \, dV + \int_B \rho_R \mathbf{b} \cdot \delta \mathbf{x} \, dV + \int_{\partial B} \mathbf{s}_0 \cdot \delta \mathbf{x} \, dS, \tag{3.66}$$

where the fields $\mathbf{b}$ and $\mathbf{s}_0$ are defined as in the preceding two subsections. It will be assumed that there are no geometric constraints on the motion of the material on ∂B_t.

The kinetic energy of the material contained in B_t is

$$T = \int_B \tfrac{1}{2} \rho_R \mathbf{v} \cdot \mathbf{v} \, dV. \tag{3.67}$$

Hamilton's principle for an arbitrary continuous medium which does not exhibit microstructural effects states: *Among comparison motions* (3.2), *the actual motion of the material is such that*

$$\int_{t_1}^{t_2} (\delta T + \delta W) \, dt = 0. \tag{3.68}$$

Since

$$\mathbf{F}^* = \mathbf{F} + \varepsilon \frac{\partial \boldsymbol{\eta}}{\partial \mathbf{X}}, \tag{3.69}$$

the variation of $\mathbf{F}$ is

$$\delta \mathbf{F} = \left[\frac{\partial \mathbf{F}^*}{\partial \varepsilon} \right]_{\varepsilon = 0} = \frac{\partial \boldsymbol{\eta}}{\partial \mathbf{X}}. \tag{3.70}$$

Using this expression and the divergence theorem, the virtual work (3.65) can be written

$$-\int_B \mathbf{S} \cdot \delta \mathbf{F} dV = -\int_{\partial B} \mathbf{S} \mathbf{N} \cdot \delta \mathbf{x} dS + \int_B \operatorname{DIV} \mathbf{S} \cdot \delta \mathbf{x} dV. \tag{3.71}$$

Therefore, using the expressions (3.66), (3.67), and (3.71), Equation (3.68) can be written in the form

$$\int_{t_1}^{t_2} \left[\int_B (-\rho_R \mathbf{a} + \operatorname{DIV} \mathbf{S} + \rho_R \mathbf{b}) \cdot \delta \mathbf{x} dV \right. \\ \left. + \int_{\partial B} (\mathbf{s}_0 - \mathbf{S} \mathbf{N}) \cdot \delta \mathbf{x} dS \right] dt = 0, \tag{3.72}$$

which is identical to Equation (3.43). It therefore leads to the same equation of balance of linear momentum

$$\rho_R \mathbf{a} = \operatorname{DIV} \mathbf{S} + \rho_R \mathbf{b} \quad \text{on } \overline{B} \times [t_1, t_2], \tag{3.73}$$

and the same boundary condition

$$\mathbf{S} \mathbf{N} = \mathbf{s}_0 \quad \text{on } \partial B \times [t_1, t_2]. \tag{3.74}$$

Alternatively, by using the definition of the Cauchy stress (3.50), Equation (3.68) can be written

$$\int_{t_1}^{t_2} \left[\int_{B_t} (-\rho \mathbf{a} + \operatorname{div} \mathbf{T} + \rho \mathbf{b}) \cdot \delta \mathbf{x} dV_t \right. \\ \left. + \int_{\partial B_t} (\mathbf{t}_0 - \mathbf{T} \mathbf{n}) \cdot \delta \mathbf{x} dS_t \right] dt = 0, \tag{3.75}$$

which is identical to Equation (3.51) and leads to the same equation of balance of linear momentum

$$\rho \mathbf{a} = \operatorname{div} \mathbf{T} + \rho \mathbf{b} \quad \text{on } \overline{B}_t \times [t_1, t_2], \tag{3.76}$$

and the same boundary condition

$$\mathbf{T} \mathbf{n} = \mathbf{t}_0 \quad \text{on } \partial B_t \times [t_1, t_2]. \tag{3.77}$$

Although this statement of Hamilton's principle leads to equations of balance of linear momentum and boundary conditions which are formally identical to those that were obtained in the case of an elastic solid, the crucial difference is that in this case the linear transformations $\mathbf{S}$ and $\mathbf{T}$ are constitutive variables. Thus

Equations (3.73)–(3.77) apply to an arbitrary continuous medium, subject only to
the restriction that the work done by internal forces as the result of a motion of the
material is of the form (3.65).[1] However, $\mathbf{S}$ and $\mathbf{T}$ are no longer derivable from a
potential energy, but must be prescribed through constitutive relations.[2]

Consider an arbitrary volume of material B' contained within B (Figure 3.1).
Let the *heat flux* $\mathbf{q}$ be a constitutive vector field that is C^1 on $\overline{B} \times [t_1, t_2]$ and is
defined such that the rate at which heat is lost from the material within B'_t by
conduction is

$$\int_{\partial B'_t} \mathbf{q} \cdot \mathbf{n}\, dS_t. \tag{3.78}$$

Let the *external heat supply* s be a prescribed scalar field that is C^0 on $\overline{B} \times [t_1, t_2]$
and is defined such that the rate at which heat is added to the material within B'_t
by external sources (such as radiation) is

$$\int_{B'_t} \rho s\, dV_t. \tag{3.79}$$

The *balance of energy postulate* for the material contained within B'_t can be written
in the form (see *e.g.* Leigh [52])

$$\frac{d}{dt} \int_{B'_t} \rho e\, dV_t = \int_{B'_t} \mathbf{T} \cdot \mathbf{L}\, dV_t - \int_{\partial B'_t} \mathbf{q} \cdot \mathbf{n}\, dS_t + \int_{B'_t} \rho s\, dV_t, \tag{3.80}$$

where $\mathbf{L} = \dfrac{\partial \mathbf{v}}{\partial \mathbf{x}}$ is the velocity gradient. Since the volume B'_t is arbitrary, this
equation implies the differential equation

$$\rho \dot{e} = \mathbf{T} \cdot \mathbf{L} - \operatorname{div} \mathbf{q} + \rho s, \tag{3.81}$$

which is called the equation of *balance of energy.*

It is easy to show that

$$\int_{B'_t} \mathbf{T} \cdot \mathbf{L}\, dV_t = \int_{B'} \mathbf{S} \cdot \dot{\mathbf{F}}\, dV. \tag{3.82}$$

This term of the energy balance postulate is called the *mechanical working term.* Observe the correspondence between the form of this term and the virtual work (3.65).
It will be shown that this correspondence can be used to motivate balance of energy

---

[1] In the next section, theories will be described in which this restriction is relaxed.

[2] Note that, since $\mathbf{S}$ and $\operatorname{DIV} \mathbf{S}$ must be continuous, the form of the constitutive equation
for $\mathbf{S}$ or $\mathbf{T}$ may impose a more stringent smoothness requirement on the motion of the
material.

postulates when Hamilton's principle is used to derive more general continuum theories. Briefly, the mechanical working terms in the balance of energy postulate are deduced from the forms of the virtual work terms in Hamilton's principle.[3] This method insures that the form of the equation of balance of energy is consistent with that of the equation of balance of linear momentum.

Thermoelasticity is an example of a theory in which the equation of balance of energy (3.81) is required (see *e.g.* Nowinski [55]). Let the *absolute temperature* $\theta(\mathbf{X}, t)$ be defined to be a non-negative scalar field that is C^2 on $\overline{B} \times [t_1, t_2]$. A thermoelastic material can be characterized by the constitutive relations ([55], Chapter 4)

$$
\begin{aligned}
\mathbf{T} &= \mathbf{T}(\mathbf{F}, \theta, \operatorname{grad} \theta), \\
e &= e(\mathbf{F}, \theta, \operatorname{grad} \theta), \\
\mathbf{q} &= \mathbf{q}(\mathbf{F}, \theta, \operatorname{grad} \theta).
\end{aligned}
\tag{3.83}
$$

These constitutive relations together with Equations (2.58), (3.76), and (3.81) provide a system of equations to determine the density ρ, the displacement $\mathbf{u}$, the temperature θ, the Cauchy stress $\mathbf{T}$, the internal energy e, and the heat flux $\mathbf{q}$.

3.2   Theories with Microstructure

In the classical theories of fluid and solid mechanics, the mechanical behavior of the material is completely described by its motion (3.1). In a continuum theory with microstructure, new fields are introduced which are independent of the motion and which describe properties of the material which the classical theories are unable to express. Hamilton's principle is a useful technique for determining the equations which govern the new fields. In this work, two specific examples of theories of this type will be described; a theory of granular solids developed by Goodman and Cowin, and a general theory of elastic solids with microstructure due to Mindlin.

When postulates of Hamilton's principle were originally formulated for the classical theories of fluid and solid mechanics, the results being sought were well known. The two examples presented in this section, and the material on mixtures in the next chapter, show how natural extensions of those postulates can be used to obtain new theories.

---

[3]See Drumheller and Bedford [22]. Similar procedures have been suggested by Ericksen [26] and Serrin ([65], p. 148).

3.2.1 Granular Solids

The work of Goodman and Cowin [32] provides an opportunity for an interesting and informative example of the use of Hamilton's principle to derive a relatively simple model of a material with microstructure. Although they did not use Hamilton's principle in developing their theory, it will be shown that it provides a natural and advantageous approach to theories of this type.

They proposed a continuum theory for application to materials consisting of solid grains with interstitial voids. They introduced a field $\phi(\mathbf{X}, t)$, the *volume fraction* of the material, which is a measure of the volume occupied by the grains per unit volume of the material. They recognized that ϕ can vary independently of the motion (3.1) as a result of deformations and reorientations of the grains. Let it be assumed that the motion (3.1) and the comparison motion (3.2) are C^3 and that the volume fraction $\phi(\mathbf{X}, t)$ is C^2 on $\overline{B} \times [t_1, t_2]$. A *comparison volume fraction field* will be defined by

$$\phi^* = \phi(\mathbf{X}, t) + \varepsilon f(\mathbf{X}, t), \tag{3.84}$$

where $f(\mathbf{X}, t)$ is an arbitrary C^2 scalar field on $\overline{B} \times [t_1, t_2]$ which satisfies the conditions $f(\mathbf{X}, t_1) = 0$ and $f(\mathbf{X}, t_2) = 0$.

The virtual work done by internal forces is postulated to be

$$\int_B [-\mathbf{S} \cdot \delta \mathbf{F} + \rho_R g \delta \phi - \mathbf{c} \cdot \delta(\,\mathrm{GRAD}\,\phi)] dV. \tag{3.85}$$

Thus the virtual work (3.65) associated with an ordinary continuous medium is supplemented by two new terms which state that work is done when changes occur in the volume fraction and in the gradient of the volume fraction of the material. It will be assumed that the Piola-Kirchhoff stress $\mathbf{S}$ and $\mathrm{DIV}\,\mathbf{S}$ are continuous on $\overline{B} \times [t_1, t_2]$. The scalar field g is a constitutive function that is assumed to be C^0 on $\overline{B} \times [t_1, t_2]$ and is called the *intrinsic equilibrated body force*. The vector field $\mathbf{c}$ is also a constitutive function, and is assumed to be C^1 on $\overline{B} \times [t_1, t_2]$.

The virtual work done by external forces that are distributed over the volume B_t is assumed to be of the form

$$\int_{B_t} (\rho \mathbf{b} \cdot \delta \mathbf{x} + \rho l \delta \phi) dV_t. \tag{3.86}$$

The prescribed body force $\mathbf{b}$ and the scalar field l are assumed to be continuous on $\overline{B} \times [t_1, t_2]$. The field l is a prescribed function called the *external equilibrated body force*.

It will be assumed that the motion of the material and the volume fraction are prescribed on the surface ∂B_1. The virtual work done by forces distributed on ∂B_2 is postulated in the form

$$\int_{\partial B_{t2}} (\mathbf{t}_0 \cdot \delta\mathbf{x} + H_0 \delta\phi) dS. \tag{3.87}$$

The prescribed external traction $\mathbf{t}_0$ and the prescribed scalar function H_0 are assumed to be continuous in time and piecewise regular on $\partial B_{t2} \times [t_1, t_2]$.

The total virtual work done on the material contained in B_t is therefore

$$\begin{aligned}
\delta W = {} & \int_B [-\mathbf{S} \cdot \delta\mathbf{F} + \rho_R g \delta\phi - \mathbf{c} \cdot \delta(\,\mathrm{GRAD}\ \phi)] dV \\[2mm]
& + \int_{B_t} (\rho\mathbf{b} \cdot \delta\mathbf{x} + \rho l \delta\phi) dV_t \\[2mm]
& + \int_{\partial B_2} (\mathbf{t}_0 \cdot \delta\mathbf{x} + H_0 \delta\phi) dS.
\end{aligned} \tag{3.88}$$

The kinetic energy of the material is written in the form

$$T = \int_{B_t} \tfrac{1}{2}\rho(\mathbf{v} \cdot \mathbf{v} + k\dot\phi^2) dV_t. \tag{3.89}$$

Thus an additional kinetic energy expression is introduced which contains the square of the material derivative of the new independent field, the volume fraction. This term is the kinetic energy associated with the local expansion and contraction of the grains, which can occur independently of the motion (3.1). In general, the coefficient k must be treated as a constitutive function.[4] For simplicity in this discussion, k will be assumed to be a constant.

Observe that the virtual work and kinetic energy expressions that have been defined follow in a natural and systematic way once the new independent field, the volume fraction, is introduced. In addition to the kinetic energy due to the translational motion of the material, a new kinetic energy expressed in terms of the rate of change of the new scalar field is included. Similarly, it is assumed that work is done when the new scalar field changes and when its gradient changes.

Hamilton's principle for a Goodman-Cowin material states: *Among comparison motions (3.2) and comparison volume fraction fields (3.84), the actual fields are such that*

$$\int_{t_1}^{t_2} (\delta T + \delta W) dt = 0. \tag{3.90}$$

---

[4]See the related discussion in Section 4.4.

Note from (3.84) that

$$\text{GRAD } \phi^* = \frac{\partial \phi}{\partial \mathbf{X}} + \varepsilon \frac{\partial f}{\partial \mathbf{X}}, \tag{3.91}$$

so that

$$\delta(\text{ GRAD } \phi) = \frac{\partial f}{\partial \mathbf{X}}. \tag{3.92}$$

Therefore the third term in the first integral of the virtual work expression (3.88) can be written

$$\int_B \mathbf{c} \cdot \delta(\text{ GRAD } \phi) dV = \int_B \mathbf{c} \cdot \frac{\partial f}{\partial \mathbf{X}} dV = \int_{B_t} \mathbf{h} \cdot \frac{\partial f}{\partial \mathbf{x}} dV_t, \tag{3.93}$$

where the vector field

$$\mathbf{h} = \frac{1}{J} \mathbf{F}\mathbf{c} \tag{3.94}$$

is called the *equilibrated stress vector*. By applying the divergence theorem, (3.93) can be expressed as

$$\int_B \mathbf{c} \cdot \delta(\text{ GRAD } \phi) dV = \int_{\partial B_2} \mathbf{h} \cdot \mathbf{n} \delta \phi \, dS - \int_{B_t} (\text{ div } \mathbf{h}) \delta \phi \, dV_t, \tag{3.95}$$

where $\delta \phi = f$.

The integral from t_1 to t_2 of the second term in the kinetic energy expression (3.89) is

$$T_2 = \int_{t_1}^{t_2} \int_{B_t} \tfrac{1}{2} \rho k \dot{\phi}^2 \, dV_t \, dt = \int_{t_1}^{t_2} \int_B \tfrac{1}{2} \rho_R k \dot{\phi}^2 \, dV \, dt. \tag{3.96}$$

In terms of the comparison volume fraction field (3.84), this is

$$T_2^* = \int_{t_1}^{t_2} \int_B \tfrac{1}{2} \rho_R k (\dot{\phi}^*)^2 \, dV \, dt. \tag{3.97}$$

The derivative of this expression with respect to ε is

$$\begin{aligned}
\frac{dT_2^*}{d\varepsilon} &= \int_{t_1}^{t_2} \int_B \rho_R k \dot{\phi}^* \frac{\partial \dot{\phi}^*}{\partial \varepsilon} \, dV \, dt \\
&= \int_{t_1}^{t_2} \int_B \rho_R k \dot{\phi}^* \dot{f} \, dV \, dt
\end{aligned} \tag{3.98}$$

Integrating this equation by parts with respect to time and evaluating the result when $\varepsilon = 0$, one obtains

$$\begin{aligned}
\delta T_2 &= -\int_{t_1}^{t_2} \int_B \rho_R k \ddot{\phi} \delta \phi \, dV \, dt \\
&= -\int_{t_1}^{t_2} \int_{B_t} \rho k \ddot{\phi} \delta \phi \, dV_t \, dt.
\end{aligned} \tag{3.99}$$

With the use of Equations (3.95) and (3.99), and by expressing the Piola-Kirchhoff stress $\mathbf{S}$ in terms of the Cauchy stress $\mathbf{T}$ through Equation (3.50), Equation (3.90) can be written

$$\int_{t_1}^{t_2} \Big[\int_{B_t} (-\rho\mathbf{a} + \operatorname{div}\mathbf{T} + \rho\mathbf{b}) \cdot \delta\mathbf{x}\,dV_t$$

$$+ \int_{B_t} (-\rho k\ddot{\phi} + \operatorname{div}\mathbf{h} + \rho l + \rho g)\delta\phi\,dV_t$$

$$+ \int_{\partial B_{t2}} (\mathbf{t}_0 - \mathbf{Tn}) \cdot \delta\mathbf{x}\,dS_t \tag{3.100}$$

$$+ \int_{\partial B_{t2}} (H_0 - \mathbf{h}\cdot\mathbf{n})\delta\phi\,dS_t \Big] dt = 0.$$

As a result of the independence of the fields $\delta\mathbf{x}$ and $\delta\phi$, Lemmas 3 and 4 of Section 2.4 can be applied to obtain the differential equations

$$\left.\begin{aligned} \rho\mathbf{a} &= \operatorname{div}\mathbf{T} + \rho\mathbf{b} \\[2mm] \rho k\ddot{\phi} &= \operatorname{div}\mathbf{h} + \rho l + \rho g \end{aligned}\right\} \quad \text{on } \overline{B}_t \times [t_1, t_2], \tag{3.101}$$

and the boundary conditions

$$\left.\begin{aligned} \mathbf{Tn} &= \mathbf{t}_0 \\[2mm] \mathbf{h}\cdot\mathbf{n} &= H_0 \end{aligned}\right\} \quad \text{on } \partial B_{t2} \times [t_1, t_2]. \tag{3.102}$$

When constitutive relations are specified for the Cauchy stress $\mathbf{T}$, the equilibrated stress vector $\mathbf{h}$, and the intrinsic equilibrated body force g, Equations (2.58) and (3.101) can be used to determine the density field ρ, the velocity field $\mathbf{v}$, and the volume fraction field ϕ.

Equations (3.101) are identical (with minor changes in notation) to the equations obtained by Goodman and Cowin ([32], Equations (4.7) and (4.10)). Although they did not use Hamilton's principle to obtain these equations, they did use a variational analysis of the static case to motivate them. The variational analysis has been described by Cowin and Goodman [17].

This theory has been used to analyze the flow of granular materials by Cowin [16], Nunziato *et al.* [57], and Passman *et al.* [63], and has been used in the study of the propagation of waves in granular materials by Cowin and Nunziato [18] and Nunziato and Walsh [58].

Two general observations can be made which are illustrated by this example. First, Hamilton's principle yields an equation for each independent field required to describe the mechanical state of a material. In this example, the fields are the motion and the volume fraction. This characteristic of Hamilton's principle makes it particularly advantageous for application to microstructure theories. Second, the generalized forces which are introduced into Hamilton's principle as virtual work terms must either be prescribed or must be specified through constitutive relations. In this example, $\mathbf{b}$, l, $\mathbf{t}_0$ and H_0 are prescribed, while $\mathbf{T}$, g and $\mathbf{h}$ are constitutive variables.

The approach described at the end of Subsection 3.1.3 can be used to postulate the equation of balance of energy for the medium. Recall the correspondence between the virtual work (3.65) containing the Piola-Kirchhoff stress and the mechanical working term (3.82) which appears in the balance of energy postulate for an ordinary material. The virtual work done by internal forces in the case of a Goodman-Cowin material is given by the expression (3.85). By analogy, the mechanical working term for an arbitrary volume B' of Goodman-Cowin material is

$$\int_{B'} [\mathbf{S} \cdot \dot{\mathbf{F}} - \rho_R g \dot{\phi} + \mathbf{c} \cdot \text{GRAD } \dot{\phi}] dV.$$

Equating this expression to the rate of change of the internal energy of the material within B' and introducing the heat conduction terms (3.78) and (3.79), a postulate of balance of energy for a Goodman-Cowin material is

$$\begin{aligned}
\frac{d}{dt} \int_{B'} \rho_R e \, dV = \ & \int_{B'} \mathbf{S} \cdot \dot{\mathbf{F}} dV - \int_{B'_t} \rho g \dot{\phi} dV_t \\
& + \int_{B'} \mathbf{c} \cdot \text{GRAD } \dot{\phi} dV - \int_{\partial B'_t} \mathbf{q} \cdot \mathbf{n} dS_t \qquad (3.103) \\
& + \int_{B'_t} \rho s \, dV_t.
\end{aligned}$$

The resulting local form of the equation of balance of energy is

$$\rho \dot{e} = \mathbf{T} \cdot \mathbf{L} - \rho g \dot{\phi} + \mathbf{h} \cdot \text{grad } \dot{\phi} - \text{div } \mathbf{q} + \rho s. \qquad (3.104)$$

This result is identical to the postulate of Goodman and Cowin ([32], Equation (4.11)).

3.2.2 Elastic Solids with Microstructure

In the theory described in the previous subsection, a new independent scalar field, the volume fraction, was introduced which provides very limited information on the local state of deformation and orientation of the grains in a granular medium.

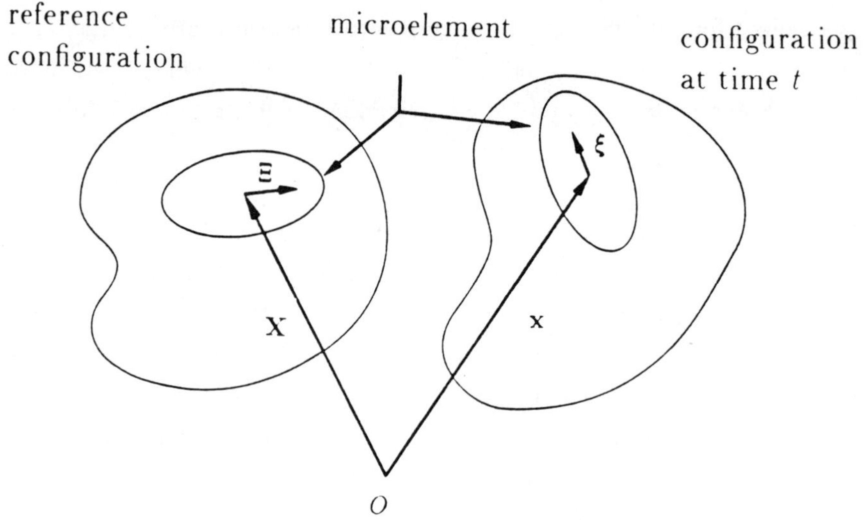

Figure 3.2: A microelement in the reference configuration and at time t.

Mindlin [54] has used Hamilton's principle to obtain a theory of linear elastic materials with microstructure which contains more extensive information concerning the local state of deformation and orientation of the material. This theory provides a clear illustration of the potential of Hamilton's principle for generating new theories of continuous media.

Mindlin associates with each point of the material a *microelement*. In the case of a granular medium, a microelement could simply represent a grain of the material. The position vector **X** of a material point in the reference configuration is assumed to be the position of the center of mass of a microelement in the reference configuration. As the result of a motion (3.1) of the material, the position of the center of mass of the microelement at time t is **x**. Let the position vector of a material point of the microelement relative to its center of mass in the reference configuration be Ξ. The position vector of the material point relative to the center of mass at time t is denoted by ξ (Figure 3.2). The *microdisplacement* of the material point of the microelement is defined by

$$\bar{\mathbf{u}} = \xi - \Xi. \tag{3.105}$$

It is then assumed that for each material point of the microelement,

$$\overline{\mathbf{u}} = \boldsymbol{\psi}^t \boldsymbol{\xi}, \tag{3.106}$$

where the linear transformation $\boldsymbol{\psi}(\mathbf{X}, t)$ is called the *microdeformation*. The microdeformation is the new independent field of the theory. It describes the state of strain of the microelement associated with each material point. Let $\boldsymbol{\psi}(\mathbf{X}, t)$ be C^2 on $\overline{B} \times [t_1, t_2]$, and define the *microdeformation comparison field* by

$$\boldsymbol{\psi}^* = \boldsymbol{\psi}(\mathbf{X}, t) + \varepsilon \mathbf{R}(\mathbf{X}, t), \tag{3.107}$$

where the linear transformation $\mathbf{R}$ is an arbitrary C^2 field on $\overline{B} \times [t_1, t_2]$ subject to the conditions that $\mathbf{R}(\mathbf{X}, t_1) = \mathbf{0}$ and $\mathbf{R}(\mathbf{X}, t_2) = \mathbf{0}$.

The strain measures of the theory are the linear strain $\mathbf{E}$, the *relative deformation*

$$\boldsymbol{\gamma} = \left(\frac{\partial \mathbf{u}}{\partial \mathbf{X}} \right)^t - \boldsymbol{\psi}, \quad \gamma_{km} = \frac{\partial u_m}{\partial X_k} - \psi_{km}, \tag{3.108}$$

and the *microdeformation gradient*

$$\boldsymbol{\kappa} = \frac{\partial \boldsymbol{\psi}}{\partial \mathbf{X}}, \quad \kappa_{kmn} = \frac{\partial \psi_{km}}{\partial X_n}. \tag{3.109}$$

Since Mindlin sought a theory for an *elastic* material with microstructure, he introduced an internal energy which is a function of the strain measures $\mathbf{E}$, $\boldsymbol{\gamma}$, and $\boldsymbol{\kappa}$. The total potential energy of the material contained in B_t is written

$$U = \int_B \rho_R e(\mathbf{E}, \boldsymbol{\gamma}, \boldsymbol{\kappa}) \, dV. \tag{3.110}$$

It will be assumed that the derivatives

$$\boldsymbol{\tau} = \rho_R \frac{\partial e}{\partial \mathbf{E}}, \quad \tau_{km} = \rho_R \frac{\partial e}{\partial E_{km}},$$

$$\boldsymbol{\sigma} = \rho_R \frac{\partial e}{\partial \boldsymbol{\gamma}}, \quad \sigma_{km} = \rho_R \frac{\partial e}{\partial \gamma_{km}}, \tag{3.111}$$

$$\boldsymbol{\mu} = \rho_R \frac{\partial e}{\partial \boldsymbol{\kappa}}, \quad \mu_{kmn} = \rho_R \frac{\partial e}{\partial \kappa_{kmn}},$$

exist and are continuous on a suitable open domain of their arguments, and that the fields $\mathrm{DIV}\,\boldsymbol{\tau}$, $\mathrm{DIV}\,\boldsymbol{\sigma}$, and $\mathrm{DIV}\,\boldsymbol{\mu}$, where

$$(\mathrm{DIV}\,\boldsymbol{\mu})_{km} = \frac{\partial \mu_{kmn}}{\partial X_n}, \tag{3.112}$$

are continuous on $\overline{B} \times [t_1, t_2]$.

It will be assumed that there are no geometrical constraints on the motion of the material or on the value of the microdeformation ψ on ∂B. The virtual work done by external forces is postulated in the form

$$\delta W = \int_B \rho_R(\mathbf{b} \cdot \delta \mathbf{x} + \mathbf{D} \cdot \delta \psi)dV$$
$$+ \int_{\partial B}(\mathbf{s}_0 \cdot \delta \mathbf{x} + \mathbf{M}_0 \cdot \delta \psi)dS. \tag{3.113}$$

The body force $\mathbf{b}$ and the linear transformation $\mathbf{D}$ are assumed to be prescribed and continuous on $\overline{B} \times [t_1, t_2]$. The field $\mathbf{D}$ is called the *double force* per unit mass. The surface traction $\mathbf{s}_0$ and the linear transformation $\mathbf{M}_0$ are also prescribed and are assumed to be continuous in time and piecewise regular on $\partial B \times [t_1, t_2]$. The field $\mathbf{M}_0$ is called the double force per unit area. Compare this equation to the virtual work expressions (3.86) and (3.87) which appeared in the theory of granular solids. In that case, the independent field was a scalar, the volume fraction. In the present theory, the independent field is a linear transformation.

The kinetic energy of the material is postulated in the form

$$T = \int_B \left(\tfrac{1}{2}\rho_R \mathbf{v} \cdot \mathbf{v} + \tfrac{1}{6}\rho'_R \dot{\psi} \cdot \mathbf{Q}\dot{\psi}\right) dV, \tag{3.114}$$

where

$$\dot{\psi} \cdot \mathbf{Q}\dot{\psi} = Q_{ijkm}\dot{\psi}_{ij}\dot{\psi}_{km}. \tag{3.115}$$

The scalar ρ'_R and the linear transformation $\mathbf{Q}$ are constants which are determined by the distribution of mass within the microelement in the reference configuration. The second term in this expression is the kinetic energy associated with the independent rotation and rate of deformation of the microelement. Compare this equation with (3.89), in which the additional kinetic energy term is due to an independent dilatational motion.

Hamilton's principle for a Mindlin material states: *Among comparison motions* (3.2) *and microdeformation comparison fields* (3.107), *the actual fields are such that*

$$\int_{t_1}^{t_2} \left[\delta(T - U) + \delta W\right] dt = 0. \tag{3.116}$$

In terms of the comparison motion and the comparison microdeformation field, the potential energy (3.110) is

$$U^* = \int_B \rho_R e(\mathbf{E}^*, \gamma^*, \kappa^*)dV. \tag{3.117}$$

The derivative with respect to ε is

$$\frac{\partial U^*}{\partial \varepsilon} = \int_B \left(\boldsymbol{\tau}^* \cdot \frac{\partial \mathbf{E}^*}{\partial \varepsilon} + \boldsymbol{\sigma}^* \cdot \frac{\partial \boldsymbol{\gamma}^*}{\partial \varepsilon} + \boldsymbol{\mu}^* \cdot \frac{\partial \boldsymbol{\kappa}^*}{\partial \varepsilon} \right) dV, \tag{3.118}$$

where

$$\boldsymbol{\mu}^* \cdot \frac{\partial \boldsymbol{\kappa}^*}{\partial \varepsilon} = \mu_{kmn}^* \frac{\partial \kappa_{kmn}^*}{\partial \varepsilon}. \tag{3.119}$$

The second term in (3.118) can be written

$$\int_B \boldsymbol{\sigma}^* \cdot \frac{\partial \boldsymbol{\gamma}^*}{\partial \varepsilon} dV = \int_B \left[\left(\boldsymbol{\sigma}^t \right)^* \cdot \frac{\partial \boldsymbol{\eta}}{\partial \mathbf{X}} - \boldsymbol{\sigma}^* \cdot \mathbf{R} \right] dV$$

$$= \int_{\partial B} \left(\boldsymbol{\sigma}^t \right)^* \mathbf{N} \cdot \boldsymbol{\eta} \, dS - \int_B \mathrm{DIV} \left(\boldsymbol{\sigma}^t \right)^* \cdot \boldsymbol{\eta} \, dV \tag{3.120}$$

$$- \int_B \boldsymbol{\sigma}^* \cdot \mathbf{R} \, dV.$$

By performing similar manipulations on the other terms in (3.118), the variation of
the potential energy can be expressed in the form

$$\delta U = \int_{\partial B} \boldsymbol{\tau}\mathbf{n} \cdot \delta\mathbf{x} \, dS - \int_B \mathrm{DIV}\,\boldsymbol{\tau} \cdot \delta\mathbf{x} \, dV$$

$$+ \int_{\partial B} \boldsymbol{\sigma}^t \mathbf{N} \cdot \delta\mathbf{x} \, dS - \int_B \mathrm{DIV}\,\boldsymbol{\sigma}^t \cdot \delta\mathbf{x} \, dV$$

$$- \int_B \boldsymbol{\sigma} \cdot \delta\boldsymbol{\psi} \, dV + \int_{\partial B} \boldsymbol{\mu}\mathbf{N} \cdot \delta\boldsymbol{\psi} \, dS \tag{3.121}$$

$$- \int_B \mathrm{DIV}\,\boldsymbol{\mu} \cdot \delta\boldsymbol{\psi} \, dV,$$

where $\delta\boldsymbol{\psi} = \mathbf{R}$ and

$$(\boldsymbol{\mu}\mathbf{N})_{km} = \mu_{kmn} N_n. \tag{3.122}$$

The integral from t_1 to t_2 of the second term in the kinetic energy (3.114) is

$$I = \int_{t_1}^{t_2} \int_B \tfrac{1}{6}\rho_R' \dot{\boldsymbol{\psi}} \cdot \mathbf{Q}\dot{\boldsymbol{\psi}} \, dV \, dt. \tag{3.123}$$

Expressing this equation in terms of the comparison microdeformation field and
taking the derivative with respect to ε yields

$$\frac{dI^*}{d\varepsilon} = \int_{t_1}^{t_2} \int_B \tfrac{1}{3}\rho_R' \mathbf{Q}\dot{\boldsymbol{\psi}}^* \cdot \dot{\mathbf{R}} \, dV \, dt. \tag{3.124}$$

This equation can then be integrated by parts to obtain

$$\delta I = - \int_{t_1}^{t_2} \int_B \tfrac{1}{3}\rho_R' \mathbf{Q}\ddot{\boldsymbol{\psi}} \cdot \mathbf{R} \, dV \, dt. \tag{3.125}$$

Using this equation, (3.113), and (3.121), Equation (3.116) can be written

$$\int_{t_1}^{t_2} \left\{ \int_B [-\rho_R \mathbf{a} + \mathrm{DIV}\,(\boldsymbol{\tau} + \boldsymbol{\sigma}^t) + \rho_R \mathbf{b}] \cdot \delta\mathbf{x}\,dV \right.$$

$$+ \int_B \left(-\tfrac{1}{3}\rho_R' \mathbf{Q}\ddot{\boldsymbol{\psi}} + \boldsymbol{\sigma} + \mathrm{DIV}\,\boldsymbol{\mu} + \rho_R \mathbf{D}\right) \cdot \delta\boldsymbol{\psi}\,dV$$

$$+ \int_{\partial B} [\mathbf{s}_0 - (\boldsymbol{\tau} + \boldsymbol{\sigma}^t)\mathbf{N}] \cdot \delta\mathbf{x}\,dS \tag{3.126}$$

$$\left. + \int_{\partial B} (\mathbf{M}_0 + \boldsymbol{\mu}\mathbf{N}) \cdot \delta\boldsymbol{\psi}\,dS \right\} dt = 0.$$

As a result of the independence of the fields $\delta\mathbf{x}$ and $\delta\boldsymbol{\psi}$, this equation yields the differential equations

$$\left. \begin{aligned} \rho_R \mathbf{a} &= \mathrm{DIV}\,(\boldsymbol{\tau} + \boldsymbol{\sigma}^t) + \rho_R \mathbf{b} \\[2ex] \tfrac{1}{3}\rho_R' \mathbf{Q}\ddot{\boldsymbol{\psi}} &= \boldsymbol{\sigma} + \mathrm{DIV}\,\boldsymbol{\mu} + \rho_R \mathbf{D} \end{aligned} \right\} \quad \text{on } \overline{B} \times [t_1, t_2]. \tag{3.127}$$

and the boundary conditions

$$\left. \begin{aligned} (\boldsymbol{\tau} + \boldsymbol{\sigma}^t)\mathbf{N} &= \mathbf{s}_0 \\[2ex] \boldsymbol{\mu}\mathbf{N} &= \mathbf{M}_0 \end{aligned} \right\} \quad \text{on } \partial B \times [t_1, t_2]. \tag{3.128}$$

Thus, in addition to the equation of balance of linear momentum, Hamilton's principle leads to an equation of motion for the microdeformation $\boldsymbol{\psi}$. Note that the two equations of motion are coupled through the term $\boldsymbol{\sigma}$. The fact that this coupling appears explicitly is one of the strengths of Hamilton's principle. It would be difficult to simply postulate Equations (3.127).

When the constitutive relation for the internal energy is specified, Equations (3.111) and (3.127) can be used to determine the fields $\boldsymbol{\tau}$, $\boldsymbol{\sigma}$, and $\boldsymbol{\mu}$, the displacement field $\mathbf{u}$, and the microdeformation field $\boldsymbol{\psi}$. Mindlin [54] obtained a linear theory by expressing the internal energy as a second order expansion in its arguments. He also used the theory to analyze the propagation of harmonic waves.

4 Mechanics of mixtures

Blood is a mixture of a liquid, plasma, and particles, primarily erythrocytes, or red cells. In an *erythrocyte sedimentation test*, a vertical tube of anticoagulated blood is allowed to stand at rest. The cells, being slightly denser than the plasma, settle to the bottom of the tube. The rate at which the upper cell boundary falls is a standard clinical test for disease. When a leak occurs in the cooling system of a nuclear reactor (known as a loss of coolant accident, or LOCA), vapor bubbles appear in the suddenly depressurized coolant fluid, and the bubbly liquid flows rapidly toward the leak. It was the study of these two very disparate problems which led to the applications of Hamilton's principle to the continuum theory of mixtures that are described in this chapter.

When the volume fraction of one constituent of a binary mixture (*i.e.* the volume occupied by that constituent per unit volume of the mixture) changes, the volume fraction of the other constituent must adjust accordingly. This *volume fraction constraint* can be introduced into a postulate of Hamilton's principle for the mixture by using the method of Lagrange multipliers. When a bubble of gas in a liquid expands or contracts, it induces a radial motion in the surrounding liquid. The inertia associated with this radial motion can be introduced into Hamilton's principle by including, in addition to the kinetic energy of translational motion of the constituents, a kinetic energy which is expressed in terms of the rate of change of the density of the gas. These ideas suggested that Hamilton's principle could be a useful method for deriving theories of mixtures.

4.1 Motions and Comparison Motions of a Mixture

4.1.1 Motions

Consider a mixture of two constituents (a binary mixture), such as a fluid containing a distribution of particles or bubbles, or a porous solid material saturated by a fluid. In general, the two constituents of the mixture can flow relative to one another. Their individual motions can be described by modeling the constituents as two superimposed continuous media. Let the symbol C_ξ denote the ξth constituent. A *motion* of C_ξ is the vector field

$$\mathbf{x} = \chi_\xi(\mathbf{X}_\xi, t), \tag{4.1}$$

where $\mathbf{x}$ is the position vector at time t of the material point of C_ξ whose position vector is $\mathbf{X}_\xi$ in a prescribed reference configuration.[1] The *inverse motion* of C_ξ is

$$\mathbf{X}_\xi = \chi_\xi^{-1}(\mathbf{x}, t). \tag{4.2}$$

Consider a finite amount of the mixture that occupies a bounded regular region B in a prescribed reference configuration at time t_1. In general, the individual motions (4.1) would cause the constituents to occupy distinct regions $B_{\xi t}$ at time t. To prevent the constituents from moving apart during the time interval $[t_1, t_2]$, it will be assumed that the displacement of each constituent of the mixture vanishes on ∂B, or, in the case of an ideal fluid constituent, it will be assumed that the normal component of the velocity vanishes on ∂B. This is equivalent to assuming the mixture to be bounded by a rigid wall.[2] As a result, the constituents occupy a single volume $B_t = B$ with a single surface $\partial B_t = \partial B$ at each time t. This assumption is not merely a theoretical convenience. At a free surface of the mixture, the constituents could actually separate as shown in Figure 4.1. Then two types of surface result; a free surface of a single constituent, and a surface that is a boundary of one constituent but not of the other. A systematic study of boundary conditions at the latter type of surface would be possible using the methods to be described in Chapter 5.

Throughout this chapter, the motions (4.1) will be assumed to be C^3 on $\overline{B} \times [t_1, t_2]$. The description of the kinematics and deformation of C_ξ in terms of its

---

[1] See the discussion of the motion of a continuous medium in Section 2.2.

[2] An alternative approach would be to express Hamilton's principle in terms of a fixed spatial volume through which the constituents are allowed to diffuse. However, Leech [51] observes that

> Many investigators ... have tried to derive, by application of Hamilton's principle, the momentum equation using the so-called Eulerian coordinate system. They have applied the principle using a fixed (control) volume. This is not Hamilton's principle, which for a continuum must be associated with a fixed aggregate or control mass.

60

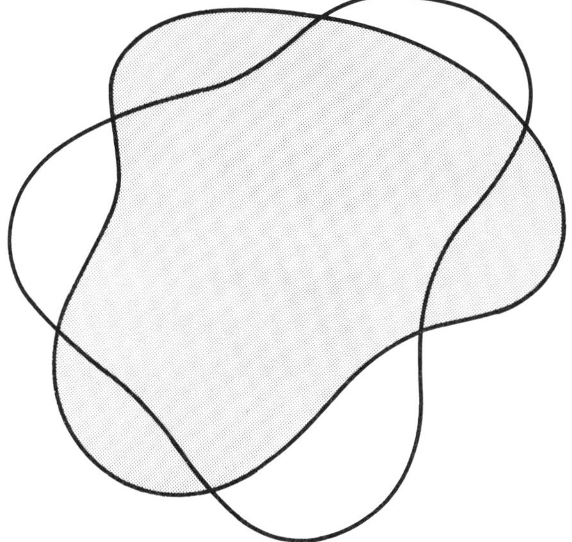

Figure 4.1: Two constituents diffusing relative to one another at the boundaries.

motion (4.1) is identical to that for a single continuous medium presented in Section 2.2. The velocity, acceleration, deformation gradient, Jacobian, displacement, and linear strain of C_ξ are defined by

$$\mathbf{v}_\xi = \frac{\partial}{\partial t}\boldsymbol{\chi}_\xi(\mathbf{X}_\xi, t),$$

$$\mathbf{a}_\xi = \frac{\partial^2}{\partial t^2}\boldsymbol{\chi}_\xi(\mathbf{X}_\xi, t),$$

$$\mathbf{F}_\xi = \frac{\partial}{\partial \mathbf{X}_\xi}\boldsymbol{\chi}_\xi(\mathbf{X}_\xi, t),$$

$$J_\xi = \det \mathbf{F}_\xi, \tag{4.3}$$

$$\mathbf{u}_\xi = \boldsymbol{\chi}_\xi(\mathbf{X}_\xi, t) - \mathbf{X}_\xi,$$

$$\mathbf{E}_\xi = \tfrac{1}{2}\left[\frac{\partial \mathbf{u}_\xi}{\partial \mathbf{X}_\xi} + \left(\frac{\partial \mathbf{u}_\xi}{\partial \mathbf{X}_\xi}\right)^t\right].$$

The material derivative of a field $\mathbf{f}_\xi(\mathbf{X}_\xi, t)$ is defined by

$$\dot{\mathbf{f}}_\xi = \frac{\partial}{\partial t}\mathbf{f}_\xi(\mathbf{X}_\xi, t) = \frac{\partial}{\partial t}\hat{\mathbf{f}}_\xi. \tag{4.4}$$

The inverse motion (4.2) maps an element dV_t of B_t at time t onto an element dV_ξ in the reference configuration. These volume elements are related by [see Equation (2.52)]

$$dV_t = J_\xi dV_\xi. \tag{4.5}$$

Let the part of dV_t occupied by C_ξ be $dV_{\xi t}$, and let the mass of C_ξ contained in $dV_{\xi t}$ be dm_ξ. The *partial density* of C_ξ is defined by $\rho_\xi = dm_\xi/dV_t$. The *material density* is defined by $\bar{\rho}_\xi = dm_\xi/dV_{\xi t}$, and the *volume fraction* is defined by $\phi_\xi = dV_{\xi t}/dV_t$. Therefore

$$\rho_\xi = \phi_\xi\bar{\rho}_\xi. \tag{4.6}$$

The fields $\rho_\xi(\mathbf{X}_\xi, t)$, $\bar{\rho}_\xi(\mathbf{X}_\xi, t)$, and $\phi_\xi(\mathbf{X}_\xi, t)$ will be assumed to be C^2 on $\overline{B} \times [t_1, t_2]$.

The equation of conservation of mass for C_ξ is

$$J_\xi = \rho_{\xi R}/\rho_\xi = \phi_{\xi R}\bar{\rho}_{\xi R}/(\phi_\xi\bar{\rho}_\xi), \tag{4.7}$$

where $\rho_{\xi R} = \rho_\xi(\mathbf{X}_\xi, t_1)$, $\phi_{\xi R} = \phi_\xi(\mathbf{X}_\xi, t_1)$, and $\bar{\rho}_{\xi R} = \bar{\rho}_\xi(\mathbf{X}_\xi, t_1)$ are the values of the partial density, volume fraction, and material density in the reference configuration. The equation of conservation of mass for C_ξ can also be expressed in the form

$$\dot{\rho}_\xi + \rho_\xi \operatorname{div} \mathbf{v}_\xi = 0. \tag{4.8}$$

In this work, consideration will be limited to mixtures for which

$$\sum_{\xi}\phi_\xi(\mathbf{x},t) = 1, \tag{4.9}$$

where $\sum_{\xi}$ means summation over the constituents of the mixture. That is, it will be assumed that the constituents of the mixture occupy all of the volume B_t at each time t; there are no voids. This equation is the *volume fraction constraint*. It plays a central role in the theories that will be discussed in this chapter.

4.1.2   Comparison Motions

A motion (4.1) of C_ξ will be called admissible if it is C^3 on $\overline{B} \times [t_1, t_2]$ and it satisfies the prescribed boundary condition on ∂B. An admissible comparison motion of C_ξ is defined in analogy with Equation (3.2),

$$\mathbf{x}_\xi^* = \boldsymbol{\chi}_\xi(\mathbf{X}_\xi, t) + \varepsilon\boldsymbol{\eta}_\xi(\mathbf{X}_\xi, t)$$

$$= \mathbf{K}_\xi(\mathbf{X}_\xi, t, \varepsilon), \tag{4.10}$$

where ε is a parameter and $\boldsymbol{\eta}_\xi(\mathbf{X}_\xi, t)$ is an arbitrary C^3 vector field on $\overline{B} \times [t_1, t_2]$ subject to the conditions that $\boldsymbol{\eta}_\xi(\mathbf{X}_\xi, t_1) = \mathbf{o}$, $\boldsymbol{\eta}_\xi(\mathbf{X}_\xi, t_2) = \mathbf{o}$, and (4.10) satisfies the prescribed boundary condition on ∂B. The inverse of the comparison motion is

$$\mathbf{X}_\xi = \mathbf{K}_\xi^{-1}(\mathbf{x}_\xi^*, t, \varepsilon). \tag{4.11}$$

This function gives the position vector in the reference configuration of the material point of C_ξ whose position vector is $\mathbf{x}_\xi^*$ at time t.

In addition, *comparison material density* and *comparison volume fraction* fields will be defined by

$$\bar{\rho}_\xi^* = \bar{\rho}_\xi(\mathbf{X}_\xi, t) + \varepsilon\bar{r}_\xi(\mathbf{X}_\xi, t),$$

$$\phi_\xi^* = \phi_\xi(\mathbf{X}_\xi, t) + \varepsilon f_\xi(\mathbf{X}_\xi, t), \tag{4.12}$$

where $\bar{r}_\xi(\mathbf{X}_\xi, t)$ and $f_\xi(\mathbf{X}_\xi, t)$ are arbitrary fields subject to the condition that they are C^2 on $\overline{B} \times [t_1, t_2]$ and that they vanish at times t_1 and t_2.

Let the volume fraction of C_ξ be expressed as a function of $\mathbf{X}_\xi, t$,

$$\phi_\xi = \hat{\phi}_\xi(\mathbf{X}_\xi, t). \tag{4.13}$$

Using this expression and the inverse motion (4.2), the volume fraction constraint (4.9) can be written

$$\sum_{\xi} \hat{\phi}_{\xi}\left(\boldsymbol{\chi}_{\xi}^{-1}(\mathbf{x},t),t\right) = 1. \tag{4.14}$$

This equation can be written in terms of the inverse of the comparison motion (4.11) and the comparison volume fraction field $(4.12)_1$ to obtain the relation

$$\sum_{\xi} \hat{\phi}_{\xi}^{*}\left(\mathbf{K}_{\xi}^{-1}(\mathbf{x},t,\varepsilon),t,\varepsilon\right) = 1. \tag{4.15}$$

The derivative of this equation with respect to ε is

$$\sum_{\xi}\left(\frac{\partial \hat{\phi}_{\xi}^{*}}{\partial \mathbf{K}_{\xi}^{-1}} \cdot \frac{\partial \mathbf{K}_{\xi}^{-1}}{\partial \varepsilon} - \frac{\partial \hat{\phi}_{\xi}^{*}}{\partial \varepsilon}\right) = 0. \tag{4.16}$$

To evaluate the partial derivative of $\mathbf{K}_{\xi}^{-1}$ with respect to ε which appears in this expression, the differential of (4.11) can be taken while holding $\mathbf{X}_{\xi}$ and t fixed to obtain

$$0 = \frac{\partial \mathbf{K}_{\xi}^{-1}}{\partial \mathbf{x}_{\xi}^{*}} d\mathbf{x}_{\xi}^{*} + \frac{\partial \mathbf{K}_{\xi}^{-1}}{\partial \varepsilon} d\varepsilon. \tag{4.17}$$

Therefore

$$\frac{\partial \mathbf{K}_{\xi}^{-1}}{\partial \varepsilon} = -\frac{\partial \mathbf{K}_{\xi}^{-1}}{\partial \mathbf{x}_{\xi}^{*}}\left[\frac{d\mathbf{x}_{\xi}^{*}}{d\varepsilon}\right]_{\mathbf{X}_{\xi},t} = -\frac{\partial \mathbf{K}_{\xi}^{-1}}{\partial \mathbf{x}_{\xi}^{*}}\boldsymbol{\eta}_{\xi}. \tag{4.18}$$

Substituting this result into (4.16) and evaluating the resulting equation when $\varepsilon = 0$, one obtains [4]

$$\sum_{\xi}(\operatorname{grad}\phi_{\xi} \cdot \delta\mathbf{x}_{\xi} - \delta\phi_{\xi}) = 0, \tag{4.19}$$

where $\delta\mathbf{x}_{\xi} = \boldsymbol{\eta}_{\xi}$ and $\delta\phi_{\xi} = f_{\xi}$. *This equation is a constraint imposed on the variations $\delta\mathbf{x}_{\xi}$ and $\delta\phi_{\xi}$ by the volume fraction constraint.* It will be introduced into statements of Hamilton's principle for mixtures in the form

$$\int_{B_t} \lambda \sum_{\xi}(\operatorname{grad}\phi_{\xi} \cdot \delta\mathbf{x}_{\xi} - \delta\phi_{\xi})dV_t = 0, \tag{4.20}$$

where the scalar field $\lambda(\mathbf{x},t)$ is a Lagrange multiplier that is assumed to be C^1 on $\overline{B}_t \times [t_1,t_2]$.

The equations of conservation of mass of the constituents (4.7) will also be introduced into statements of Hamilton's principle for mixtures as constraints, in the same form as in the case of a single continuous medium [see Equation (2.90)],

$$
\sum_\xi \int_{B_t} \pi_\xi \left(1 - \frac{\phi_{\xi R} \bar{\rho}_{\xi R}}{\phi_\xi \bar{\rho}_\xi J_\xi} \right) dV_t
$$
$$
= \sum_\xi \int_B \pi_\xi \left(J_\xi - \frac{\phi_{\xi R} \bar{\rho}_{\xi R}}{\phi_\xi \bar{\rho}_\xi} \right) dV,
\tag{4.21}
$$

where the scalar fields $\pi_\xi(\mathbf{X}_\xi, t)$ are Lagrange multipliers that are assumed to be C^1 on $\overline{B} \times [t_1, t_2]$. To determine the variation of this expression, it is written in terms of the comparison motion (4.10) and the comparison fields (4.12).

$$
\sum_\xi \int_B \pi_\xi \left(J_\xi^* - \frac{\phi_{\xi R} \bar{\rho}_{\xi R}}{\phi_\xi^* \bar{\rho}_\xi^*} \right) dV.
\tag{4.22}
$$

Taking the derivative with respect to ε and setting $\varepsilon = 0$ yields

$$
\sum_\xi \int_B \pi_\xi J_\xi \left(\operatorname{div} \boldsymbol{\eta}_\xi + \frac{f_\xi}{\phi_\xi} + \frac{\bar{r}_\xi}{\bar{\rho}_\xi} \right) dV
$$
$$
= \sum_\xi \int_{B_t} \left[- \operatorname{grad} \pi_\xi \cdot \delta \mathbf{x}_\xi + \pi_\xi \left(\frac{\delta \phi_\xi}{\phi_\xi} + \frac{\delta \bar{\rho}_\xi}{\bar{\rho}_\xi} \right) \right] dV_t,
\tag{4.23}
$$

where $\delta \bar{\rho}_\xi = \bar{r}_\xi$.

The postulates of Hamilton's principle for mixtures that will be introduced in the following sections are closely analogous to those for a single material that were described in Chapter 3. The volume fraction constraint is a new element, and new degrees of freedom will be seen to arise in comparison with the theories of single materials without microstructure. The formulations have some elements in common with the theory of granular solids that was discussed in Subsection 3.2.1.

4.2 Mixtures of Ideal Fluids

4.2.1 Compressible Fluids

As the first example, consider a binary mixture of elastic, ideal fluids. A postulate of Hamilton's principle for such a mixture will be formulated that is a direct extension of the postulate for a single elastic ideal fluid in Subsection 3.1.1.

It will be assumed that each constituent C_ξ has an internal energy per unit mass $e_\xi(\bar{\rho}_\xi)$ which is a function only of the material density of that constituent.[3] The

---

[3]This assumption, like the assumptions made in Section 3.1 that led to theories of elastic fluids and elastic solids, will obviously result in a very special theory.

second derivatives of these functions will be assumed to be continuous. The total potential energy of the mixture contained in B_t is assumed to be the sum of the potential energies of the constituents,

$$U = \sum_\xi \int_{B_t} \rho_\xi e_\xi(\overline{\rho}_\xi) dV_t. \tag{4.24}$$

The virtual work done on the mixture by external forces is postulated in the form

$$\delta W = \sum_\xi \int_{B_t} (\rho_\xi \mathbf{b}_\xi + \mathbf{d}_\xi) \cdot \delta \mathbf{x} dV_t. \tag{4.25}$$

The external force on each constituent is decomposed into two parts in this expression. The body force $\mathbf{b}_\xi$ is the force per unit mass exerted on C_ξ by external agencies, such as gravity. It is assumed to be prescribed. The *interaction force* or *drag* $\mathbf{d}_\xi$ is the force per unit volume exerted on C_ξ by the other constituent of the mixture.[4] The vector fields $\mathbf{b}_\xi(\mathbf{X}_\xi, t)$ and $\mathbf{d}_\xi(\mathbf{X}_\xi, t)$ will be assumed to be C^0 on $\overline{B} \times [t_1, t_2]$. Recall that the mixture is assumed to be bounded by a rigid wall. Therefore, no virtual work is done by external forces at the surface ∂B_t.

The comparison fields (4.10) and (4.12) are subject to the constraints (4.20) and (4.23) arising from the volume fraction constraint and the equations of conservation of mass. Therefore, the constraint term

$$\begin{aligned}
\delta C = \quad & \sum_\xi \int_{B_t} \lambda(\operatorname{grad} \phi_\xi \cdot \delta \mathbf{x}_\xi - \delta \phi_\xi) dV_t \\
+ \quad & \sum_\xi \int_{B_t} \left[-\operatorname{grad} \pi_\xi \cdot \delta \mathbf{x}_\xi + \pi_\xi \left(\frac{\delta \phi_\xi}{\phi_\xi} + \frac{\delta \overline{\rho}_\xi}{\overline{\rho}_\xi} \right) \right] dV_t
\end{aligned} \tag{4.26}$$

will be included in Hamilton's principle.

The total kinetic energy of the mixture in B_t will be assumed to be the sum of the kinetic energies due to the translational motions of the constituents,

$$T = \sum_\xi \int_{B_t} \tfrac{1}{2} \rho_\xi \mathbf{v}_\xi \cdot \mathbf{v}_\xi dV_t. \tag{4.27}$$

A postulate of Hamilton's principle for a mixture of elastic ideal fluids states [4]: *Among comparison motions (4.10) and comparison fields (4.12), the actual fields are*

---

[4]Although the constituents are here being treated as inviscid with regard to their macroscopic behavior, it is nevertheless assumed that they may exert drag forces on one another. This is a common assumption in mixture theories for processes in which macroscopic viscous effects may be neglected (see *e.g.* Bowen [12]).

66

such that

$$\int_{t_1}^{t_2} [\delta(T - U) + \delta C + \delta W]dt = 0. \tag{4.28}$$

Using the expressions (4.24)–(4.27), this equation can be written

$$\sum_\xi \int_{t_1}^{t_2} \int_{B_t} \Big[\; (-\rho_\xi \mathbf{a}_\xi + \rho_\xi \mathbf{b}_\xi + \mathbf{d}_\xi - \operatorname{grad} \pi_\xi + \lambda \operatorname{grad} \phi_\xi) \cdot \delta \mathbf{x}_\xi$$
$$+ \left(-\rho_\xi \frac{de_\xi}{d\bar\rho_\xi} + \pi_\xi/\bar\rho_\xi \right) \delta\bar\rho_\xi \tag{4.29}$$
$$+ (\pi_\xi/\phi_\xi - \lambda)\delta\phi_\xi \Big] dV_t dt = 0.$$

Because of the independence of the fields $\delta\mathbf{x}_\xi$, $\delta\bar\rho_\xi$, and $\delta\phi_\xi$ for each constituent, Lemma 3 of Section 2.4 can be invoked to obtain the differential equations

$$\left.\begin{aligned} \rho_\xi \mathbf{a}_\xi &= \rho_\xi \mathbf{b}_\xi + \mathbf{d}_\xi - \operatorname{grad} \pi_\xi + \lambda \operatorname{grad} \phi_\xi, \\[2mm] \pi_\xi &= \phi_\xi \bar\rho_\xi^2 \frac{de_\xi}{d\bar\rho_\xi}, \\[2mm] \pi_\xi &= \phi_\xi \lambda \end{aligned}\right\} \quad \text{on } \overline{B}_t \times [t_1, t_2]. \tag{4.30}$$

When constitutive relations are specified for the internal energies $e_\xi(\bar\rho_\xi)$ and the drag terms $\mathbf{d}_\xi$, these three equations together with Equations (4.8) and (4.9) can be used to determine the fields ρ_ξ, $\bar\rho_\xi$, ϕ_ξ, λ, $\mathbf{v}_\xi$, and π_ξ.

Equations (4.30) are quite revealing about the behavior of this simple model for a mixture. Note from Equation (3.17) that the term $\bar\rho_\xi^2 \dfrac{de_\xi}{d\bar\rho_\xi}$ is the pressure of C_ξ. From (4.30),

$$\lambda = \bar\rho_\xi^2 \frac{de_\xi}{d\bar\rho_\xi}. \tag{4.31}$$

Thus a consequence of this postulate of Hamilton's principle is that *the pressures of the constituents must be equal.* This condition is often introduced as an assumption in theoretical studies of multiphase flow. Models for mixtures in which this condition does not hold will be discussed later in this section. Also, observe that Equation (4.30)$_1$ can be written

$$\rho_\xi \mathbf{a}_\xi = \rho_\xi \mathbf{b}_\xi + \mathbf{d}_\xi - \phi_\xi \operatorname{grad} \lambda. \tag{4.32}$$

The form of the last term in this equation has been a subject of some controversy among those interested in theoretical models for mixtures. The form of the term which appears here is a consequence of including the volume fraction constraint in Hamilton's principle.[5]

4.2.2 Incompressible Fluids

If both constituents of the mixture are incompressible, $\overline{\rho}_\xi = \overline{\rho}_{\xi R} = $ constant, Equation (4.28) assumes the form

$$\int_{t_1}^{t_2} (\delta T + \delta W + \delta C)dt = 0. \tag{4.33}$$

The expressions for δT and δW are unchanged. The only change in δC is that $\delta \overline{\rho}_\xi = 0$. The resulting differential equations are easily shown to be

$$\left. \begin{aligned} \rho_\xi \mathbf{a}_\xi &= \rho_\xi \mathbf{b}_\xi + \mathbf{d}_\xi - \operatorname{grad} \pi_\xi + \lambda \operatorname{grad} \phi_\xi \\[2ex] \pi_\xi &= \phi_\xi \lambda \end{aligned} \right\} \quad \text{on } \overline{B}_t \times [t_1, t_2]. \tag{4.34}$$

Eliminating the Lagrange multipliers π_ξ yields the equations of balance of linear momentum

$$\rho_\xi \mathbf{a}_\xi = \rho_\xi \mathbf{b}_\xi + \mathbf{d}_\xi - \phi_\xi \operatorname{grad} \lambda. \tag{4.35}$$

For incompressible constituents, the equations of conservation of mass (4.8) can be written

$$\dot{\phi}_\xi + \phi_\xi \operatorname{div} \mathbf{v}_\xi = 0. \tag{4.36}$$

If constitutive relations are specified for the drag terms $\mathbf{d}_\xi$, Equations (4.9), (4.35), and (4.36) can be used to determine the fields λ, $\mathbf{v}_\xi$, and ϕ_ξ. The Lagrange multiplier λ is the pressure of the constituents.

The statement of Hamilton's principle for a granular solid described in Subsection 3.2.1 contained a virtual work expressed in terms of the variation of the volume fraction of the material. Would the physics of the problem justify the inclusion of such terms in a theory for a mixture of ideal fluids? Suppose that the theory is used to model a fluid containing a distribution of particles. If the particles are sufficiently small, they will undergo mutual impacts as a result of their Brownian motions. This *diffusive effect* of particle impacts is analogous to the ordinary pressure which arises in a fluid due to impacts on the molecular scale. The particles can also exert forces on one another through hydrodynamic interactions when the mixture is in motion.

---

[5]The topics mentioned in this paragraph have been discussed at length by Bedford and Drumheller [6]

68

Either of these phenomena will result in work being done when the volume fraction of the particles changes.

If a virtual work term of the form

$$-\sum_{\xi}\int_{B_t}(P_\xi/\phi_\xi)\delta\phi_\xi dV_t \tag{4.37}$$

is added to (4.25), the postulates of Hamilton's principle that have been stated are extended to the case of ideal fluids with diffusivity [48].[6] The scalar fields $P_\xi(\mathbf{X}_\xi,t)$ are assumed to be C^1 on $\overline{B}\times[t_1,t_2]$. In the case of incompressible constituents, the resulting differential equations are

$$\left.\begin{aligned}\rho_\xi\mathbf{a}_\xi &= \rho_\xi\mathbf{b}_\xi + \mathbf{d}_\xi - \operatorname{grad}\pi_\xi + \lambda\operatorname{grad}\phi_\xi \\ \pi_\xi &= \phi_\xi\lambda + P_\xi\end{aligned}\right\} \quad \text{on } \overline{B}_t\times[t_1,t_2]. \tag{4.38}$$

Eliminating the Lagrange multipliers π_ξ results in the equations of balance of linear momentum

$$\rho_\xi\mathbf{a}_\xi = \rho_\xi\mathbf{b}_\xi + \mathbf{d}_\xi - \phi_\xi\operatorname{grad}\lambda - \operatorname{grad}P_\xi. \tag{4.39}$$

In addition to the gradient of the pressure λ appearing in the equations of motion, the gradients of the *diffusive pressures* P_ξ also appear.

The constitutive relations postulated by Hill *et al.* [48] for the sedimentation of a distribution of rigid particles in an incompressible fluid are

$$\mathbf{d}_\xi = \mathbf{d}_\xi(\phi_\xi, \mathbf{v}_\xi - \mathbf{v}_\gamma),$$

$$P_\xi = P_\xi(\phi_\xi, \mathbf{v}_\xi - \mathbf{v}_\gamma), \tag{4.40}$$

where $\xi\neq\gamma$. Note that since the volume fractions are related through the volume fraction constraint, it is not necessary to assume that the constitutive relations are functions of both volume fractions. If these equations are assumed to be isotropic and linear in the relative velocity, they become [48]

$$\mathbf{d}_\xi = \alpha_\xi(\mathbf{v}_\xi - \mathbf{v}_\gamma),$$

$$P_\xi = \beta_\xi, \tag{4.41}$$

where α_ξ and β_ξ are scalar functions of ϕ_ξ.

---

[6]The term P_ξ/ϕ_ξ is the generalized force. This expression is used simply because it has been found to result in simpler equations. The justification for doing so is that the P_ξ are assumed to be constitutive terms which depend upon the volume fractions.

For purposes of comparison with this derivation, equivalent theories have been derived using at least two other approaches. Craine [19] used postulated equations of motion and introduced the volume fraction constraint into the second law of thermodynamics (the Clausius-Duhem inequality) for the mixture. Drew [20] used an averaging method.

Equations (4.9), (4.36), (4.39), and (4.41) have been applied to the erythrocyte sedimentation test described in the introduction to this chapter by Hill and Bedford [47] and Hill *et al.* [48]. Results of their numerical solutions of the equations are compared in Figures 4.2 and 4.3 to experimental measurements made using anticoagulated human whole blood by Whelan *et al.* [75]. In Figure 4.2, the predicted distribution of the cell volume fraction as a function of height in the vertical tube is compared to measurements made at several times. An empirical expression for the drag coefficient α_ξ was used, and the constitutive coefficients in the theory were chosen to obtain the best agreement with the data at 2 hours. They were then left unchanged while the computations were extended to 4 hours and 8.5 hours. In Figure 4.3, the prediction of the upper cell boundary position as a function of time, which is used as a clinical indicator of disease, is compared to the observed position.

This theory has also been used to study the stability of steady sedimentation of a uniform distribution of particles in an incompressible fluid by Hill [45] and Hill and Bedford [46].

4.2.3   Fluids with Microinertia

Suppose that there is a spherical bubble of gas in an unbounded, incompressible liquid. Let the radius of the bubble be R, and let the mass densities of the gas and the liquid be $\bar{\rho}_g$ and $\bar{\rho}_f$. If the bubble expands or contracts, it will induce a radial velocity distribution in the liquid. The velocity of the liquid at a distance r from the center of the bubble is

$$v_f = \left(\frac{R}{r}\right)^2 \dot{R}, \tag{4.42}$$

where the dot denotes the time derivative. The total kinetic energy of the liquid surrounding the bubble is[7]

$$\int_R^\infty \tfrac{1}{2}\bar{\rho}_f 4\pi r^2 v_f^2 \, dr. \tag{4.43}$$

---

[7]The kinetic energy of the expanding gas is usually neglected due to the much greater density of the liquid (see *e.g.* van Wijngaarden [72]).

70

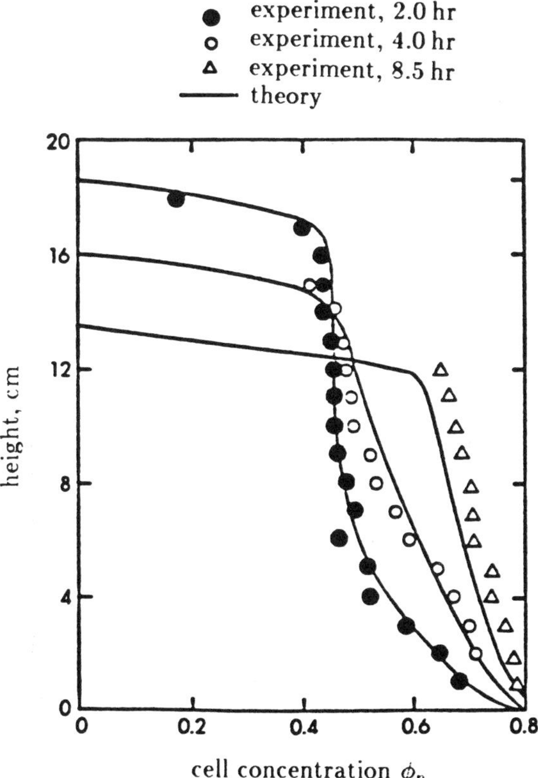

Figure 4.2: Comparison of the mixture theory with cell concentration profiles measured in blood sedimentation.

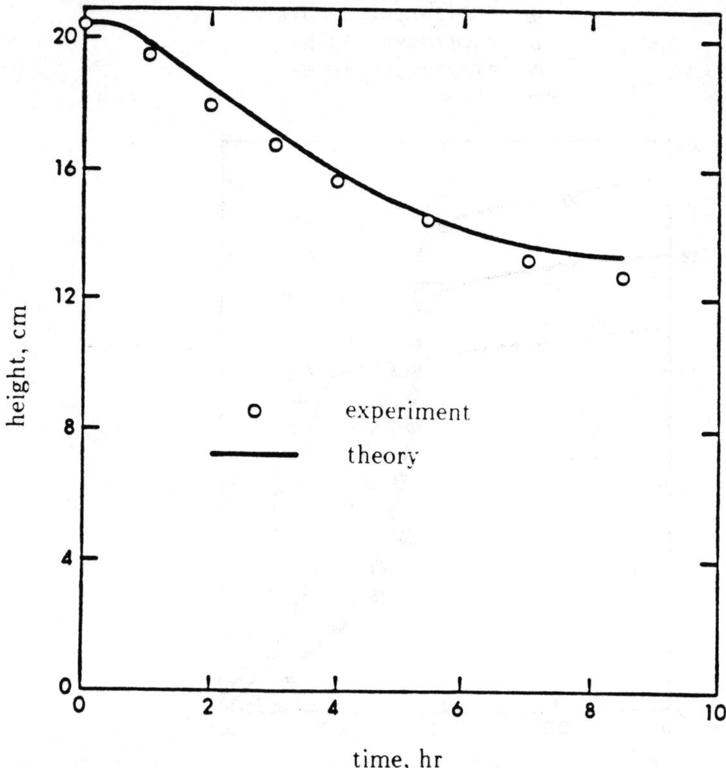

Figure 4.3: Comparison of the mixture theory with the upper cell interface position observed in blood sedimentation.

Substituting the velocity distribution (4.42), this integral can be evaluated to obtain

$$2\pi\bar{\rho}_f R^3 \dot{R}^2. \tag{4.44}$$

Since the mass of the gas in the bubble $\bar{\rho}_g \frac{4}{3}\pi R^3 = $ constant, the kinetic energy (4.44) can be written

$$\left[\frac{2\pi\bar{\rho}_f(\bar{\rho}_{gR})^{\frac{5}{3}} R_R^5}{9(\bar{\rho}_g)^{\frac{11}{3}}}\right]\dot{\bar{\rho}}_g^2, \tag{4.45}$$

where $\bar{\rho}_{gR}$ and R_R are reference values.

Now consider a liquid containing a dilute distribution of bubbles, and suppose that in a prescribed reference configuration the bubbles are uniformly distributed and are of equal radius R_R and density $\bar{\rho}_{gR}$. In a motion of this bubbly liquid, the bubbles will undergo volumetric oscillations and induce local radial motions of the liquid. If it is assumed that the kinetic energy of the liquid surrounding a single bubble can be approximated by the expression (4.45), the kinetic energy per unit volume of the mixture due to local radial motions of the liquid can be obtained by multiplying (4.45) by the number of bubbles per unit volume $\phi_g/(\frac{4}{3}\pi R^3)$, where ϕ_g is the volume fraction of the gas. The result is

$$\rho_g\left[\frac{\bar{\rho}_f(\bar{\rho}_{gR})^{\frac{2}{3}} R_R^2}{6(\bar{\rho}_g)^{\frac{11}{3}}}\right]\dot{\bar{\rho}}_g^2. \tag{4.46}$$

This *microkinetic energy* due to bubble oscillations has a dominant effect on the dynamic behavior of bubbly liquids (see van Wijngaarden [72]). The equations which govern a bubbly liquid can be obtained in a very straightforward way by including the microkinetic energy in Hamilton's principle [21].

For simplicity in this presentation, the relative motion between the bubbles and the liquid will be neglected. This is an acceptable approximation in many applications because of the relatively small mass of the bubbles. Therefore the mixture will be assumed to have a single motion (3.1). The microkinetic energy of the mixture contained in B_t will be expressed in the form

$$T_m = \int_{B_t} \tfrac{1}{2}\rho_g I_g(\bar{\rho}_\gamma)\dot{\bar{\rho}}_g^2 \, dV_t = \int_B \tfrac{1}{2}\rho_{gR} I_g(\bar{\rho}_\gamma)\dot{\bar{\rho}}_g^2 \, dV_g. \tag{4.47}$$

The term $I_g(\bar{\rho}_\gamma)$ is a constitutive coefficient which is assumed to be a function of each of the material densities. This expression for the microkinetic energy is motivated by (4.46) and includes it as a special case. The integral from t_1 to t_2 of this equation is

$$I = \int_{t_1}^{t_2} T_m dt = \int_{t_1}^{t_2} \int_B \tfrac{1}{2}\rho_{gR} I_g(\bar{\rho}_\gamma)\dot{\bar{\rho}}_g^2 dV_g dt. \qquad (4.48)$$

Proceeding in the now familiar way to determine the variation, this equation is expressed in terms of the comparison material density $(4.12)_1$ to obtain

$$I^*(\varepsilon) = \int_{t_1}^{t_2} \int_B \tfrac{1}{2}\rho_{gR} I_g(\bar{\rho}_\gamma^*)(\dot{\bar{\rho}}_g^*)^2 dV_g dt. \qquad (4.49)$$

The derivative with respect to ε is

$$\frac{dI^*(\varepsilon)}{d\varepsilon} = \int_{t_1}^{t_2} \int_B \left[\rho_{gR} I_g^* \dot{\bar{\rho}}_g^* \dot{\bar{\tau}}_g + \tfrac{1}{2}\rho_{gR} \left(\sum_\gamma \frac{\partial I_g^*}{\partial \bar{\rho}_\gamma^*}\bar{\tau}_\gamma \right) (\dot{\bar{\rho}}_g^*)^2 \right] dV_g dt, \qquad (4.50)$$

where $I_g^* = I_g(\bar{\rho}_\gamma^*)$. Integrating the first term by parts with respect to time and setting $\varepsilon = 0$ yields the variation

$$\delta T_m = \int_{B_t} \left[-\rho_g \overline{I_g \dot{\bar{\rho}}_g} \delta\bar{\rho}_g + \tfrac{1}{2}\rho_g \left(\sum_\gamma \frac{\partial I_g}{\partial \bar{\rho}_\gamma}\delta\bar{\rho}_\gamma \right) \dot{\bar{\rho}}_g^2 \right] dV_t. \qquad (4.51)$$

The total kinetic energy of the mixture contained in B_t is

$$T = \int_{B_t} [\tfrac{1}{2}\rho\mathbf{v}\cdot\mathbf{v} + \tfrac{1}{2}\rho_g I_g(\bar{\rho}_\gamma)\dot{\bar{\rho}}_g^2]dV_t, \qquad (4.52)$$

where $\rho = \rho_f + \rho_g$ is the density of the mixture.

The potential energy will be postulated to be the total internal energy for a mixture of two compressible ideal fluids,

$$U = \sum_\xi \int_{B_t} \rho_\xi e_\xi(\bar{\rho}_\xi)dV_t. \qquad (4.53)$$

Since there is no relative motion between the constituents, the virtual work done by the drag forces vanishes, and the only virtual work done is that due to the external body force,

$$\delta W = \int_{B_t} \rho\mathbf{b}\cdot\delta\mathbf{x}dV_t. \qquad (4.54)$$

The constraint arising from the equations of balance of mass and the volume fraction constraint, Equation (4.26), is altered only by the fact that there is but a single motion.

$$\delta C = \sum_\xi \int_{B_t} \lambda(\operatorname{grad}\phi_\xi\cdot\delta\mathbf{x} - \delta\phi_\xi)dV_t$$
$$+ \sum_\xi \int_{B_t} \left[-\operatorname{grad}\pi_\xi\cdot\delta\mathbf{x} + \pi_\xi\left(\frac{\delta\phi_\xi}{\phi_\xi} + \frac{\delta\bar{\rho}_\xi}{\bar{\rho}_\xi}\right) \right] dV_t \qquad (4.55)$$

Based on (4.52)–(4.55), a statement of Hamilton's principle for an ideal compressible liquid containing a distribution of bubbles of an ideal gas [23] is: *Among comparison motions (3.2) and comparison fields (4.12), the actual fields are such that*

$$\int_{t_1}^{t_2} \left[\delta(T - U) + \delta C + \delta W\right] dt = 0. \tag{4.56}$$

Using the result (4.51), this equation can be written

$$\int_{t_1}^{t_2} \int_{B_t} \Bigg\{ \left[-\rho \mathbf{a} + \rho \mathbf{b} - \operatorname{grad}\left(\pi_f + \pi_g\right)\right] \cdot \delta \mathbf{x}$$
$$+ \left(-\rho_g \overline{I_g \dot{\bar{\rho}}_g} + \tfrac{1}{2}\rho_g \frac{\partial I_g}{\partial \bar{\rho}_g} \dot{\bar{\rho}}_g^2 - \rho_g \frac{de_g}{d\bar{\rho}_g} + \pi_g/\bar{\rho}_g\right) \delta \bar{\rho}_g$$
$$+ \left(\tfrac{1}{2}\rho_g \frac{\partial I_g}{\partial \bar{\rho}_f} \dot{\bar{\rho}}_g^2 - \rho_f \frac{de_f}{d\bar{\rho}_f} + \pi_f/\bar{\rho}_f\right) \delta \bar{\rho}_f$$
$$+ (\pi_g/\phi_g - \lambda)\delta\phi_g + (\pi_f/\phi_f - \lambda)\delta\phi_f \Bigg\} dV_t dt = 0, \tag{4.57}$$

which yields the differential equations

$$\left. \begin{array}{r}
\rho \mathbf{a} = \rho \mathbf{b} - \operatorname{grad}\left(\pi_f + \pi_g\right) \\[2ex]
\rho_g \overline{I_g \dot{\bar{\rho}}_g} - \tfrac{1}{2}\rho_g \dfrac{\partial I_g}{\partial \bar{\rho}_g} \dot{\bar{\rho}}_g^2 = -\rho_g \dfrac{de_g}{d\bar{\rho}_g} + \pi_g/\bar{\rho}_g \\[2ex]
- \tfrac{1}{2}\rho_g \dfrac{\partial I_g}{\partial \bar{\rho}_f} \dot{\bar{\rho}}_g^2 = -\rho_f \dfrac{de_f}{d\bar{\rho}_f} + \pi_f/\bar{\rho}_f \\[2ex]
\pi_g = \phi_g \lambda \\[2ex]
\pi_f = \phi_f \lambda
\end{array} \right\} \quad \text{on } \overline{B}_t \times [t_1, t_2]. \tag{4.58}$$

The last two equations can be used to eliminate the Lagrange multipliers π_f and π_g, resulting in the three equations

$$\rho \mathbf{a} = \rho \mathbf{b} - \operatorname{grad} \lambda,$$

$$\bar{\rho}_g^2 \overline{I_g \dot{\bar{\rho}}_g} - \tfrac{1}{2}\bar{\rho}_g \left(\bar{\rho}_g \frac{\partial I_g}{\partial \bar{\rho}_g} - \frac{\phi_g}{\phi_f}\bar{\rho}_f \frac{\partial I_g}{\partial \bar{\rho}_f}\right) \dot{\bar{\rho}}_g^2 = p_f - p_g, \tag{4.59}$$

$$\lambda = p_f - \tfrac{1}{2}\bar{\rho}_g \frac{\phi_g}{\phi_f}\bar{\rho}_f \frac{\partial I_g}{\partial \bar{\rho}_f} \dot{\bar{\rho}}_g^2,$$

where the constituent pressures p_ξ are

$$p_\xi = \bar{\rho}_\xi^2 \frac{de_\xi}{d\bar{\rho}_\xi}. \tag{4.60}$$

When constitutive equations are specified for the internal energies e_ξ and the coefficient I_g, the equations of conservation of mass

$$\dot{\rho}_\xi + \rho_\xi \operatorname{div} \mathbf{v} = 0 \qquad (4.61)$$

together with Equations (4.9), (4.59) and (4.60) provide a system of equations with which to determine the fields ϕ_ξ, λ, $\mathbf{v}$, $\bar{\rho}_\xi$, and p_ξ.

Observe from Equation $(4.59)_2$ that, as a result of introducing the microkinetic energy, the pressures of the constituents are not in general equal. It is the difference in the liquid and gas pressures that drives the bubble oscillations. The pressures are equal when the mixture is in a state of equilibrium. Also, note that the term λ which appears in the equation of balance of linear momentum $(4.59)_1$ is *not* in general equal to the pressure of the liquid.

These equations have been compared to experimental data on wave propagation in bubbly liquids by Bedford and Stern [7] and Drumheller *et al.* [23]. In order to do so, it was necessary to account for the effects of heat transfer from the gas to the liquid in determining the constitutive equation for the gas pressure (see *e.g.* Drumheller and Bedford [21]). The coefficient I_g was evaluated using the expression (4.46). This seems contradictory since the liquid is assumed to be compressible while (4.46) was derived under the assumption that it is incompressible. The assumption that is necessary in order to use this expression to approximate I_g is that the spatial variation of the density of the liquid is small in the neighborhood of a bubble. That is, wavelengths must be large in comparison to the bubble diameter.

In Figures 4.4 and 4.5, the predicted phase velocity and attenuation of plane acoustic waves are compared to measurements by Silberman [66] for air bubbles in water. The gas volume fraction was $\phi_{gR} = 3.77 \times 10^{-4}$ and the bubble radius was $1.01\,\mathrm{mm}$. The peak in the attenuation occurs at the resonance frequency for bubble oscillations.

In Figure 4.6, a numerical solution for the gas pressure p_g resulting from an impulsively applied pressure is compared to data obtained with a shock tube by Kuznetsov *et al.* [49]. The mixture consisted of carbon dioxide bubbles in a water-glycerine solution. The gas volume fraction was $\phi_{gR} = 0.01$ and the bubble radius was $0.5\,\mathrm{mm}$. The "ringing" observed in the pressure history results from bubble oscillations.

If the liquid is assumed to be incompressible, the coefficient I_g is evaluated using (4.46), and Equation $(4.59)_2$ is expressed in terms of the bubble radius R

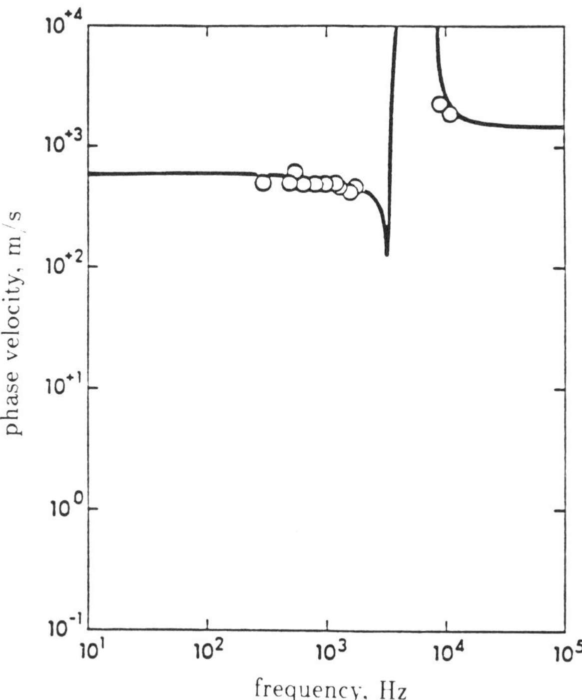

Figure 4.4: Phase velocity of acoustic waves in a water-air mixture.

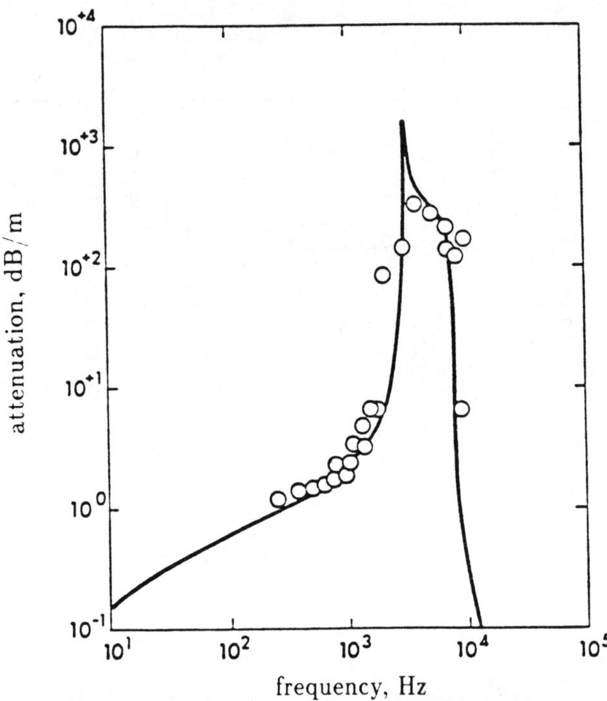

Figure 4.5: Attenuation of acoustic waves in a water-air mixture.

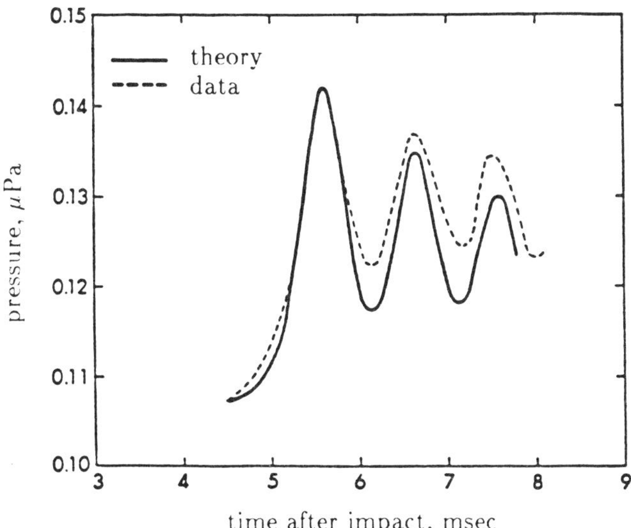

Figure 4.6: Transient pressure history in a bubbly liquid compared with shock tube data.

instead of $\bar{\rho}_g$, Equations (4.59) become

$$\rho \mathbf{a} = \rho \mathbf{b} - \operatorname{grad} \lambda,$$

$$R\ddot{R} + \frac{3}{2}\left(1 - \frac{\phi_g}{\phi_f}\right)\dot{R}^2 = (p_g - p_f)/\bar{\rho}_f, \qquad (4.62)$$

$$\lambda = p_f - \frac{3}{2}\frac{\phi_g}{\phi_f}\bar{\rho}_f\dot{R}^2.$$

In the limit $\phi_g \to 0$, these equations reduce to

$$\rho \mathbf{a} = \rho \mathbf{b} - \operatorname{grad} p_f,$$

$$R\ddot{R} + \frac{3}{2}\dot{R}^2 = (p_g - p_f)/\bar{\rho}_f. \qquad (4.63)$$

Equation $(4.63)_2$ is the *Rayleigh-Plesset equation* for the dilatational motion of a single bubble in an unbounded liquid. That this equation is obtained from a mixture theory of this type has been pointed out by Drumheller *et al.* [23] and independently by Passman *et al.* [64].

The approach which is usually used to model the dynamics of bubbly liquids is to adopt Equation $(4.63)_1$, which is simply the equation of balance of linear momentum for an ideal fluid, and to assume that $(4.63)_2$ applies [72]. Thus this model is recovered from the equations derived through Hamilton's principle in the limit as the bubble volume fraction approaches zero.

Hamilton's principle has been used to obtain the governing equations for a bubbly liquid in which there is relative motion between the liquid and the bubbles by Drumheller and Bedford [21].

4.3 Mixture of an Ideal Fluid and an Elastic Solid

The application of Hamilton's principle to a binary mixture of an elastic, ideal fluid and an elastic solid is discussed in this section. Although this case requires only a minor extension of the formulation for mixtures of ideal fluids, it provides an introduction to the more general theory that is described in the next section. Another reason that this case deserves attention is that it leads to Biot's theory for a fluid saturated porous elastic material, which is the most widely accepted and applied theory of mixtures.

Let the fluid and solid constituents be denoted by subscripts f and s. The potential energy of the mixture contained in B_t will be expressed in the form

$$U = \int_{B_t} [\rho_f e_f(\overline{\rho}_f) + \rho_s e_s(\overline{\rho}_s, \mathbf{E}_s)] dV_t, \tag{4.64}$$

where $\mathbf{E}_s$ is the linear strain of the elastic material. The material density $\overline{\rho}_s$ of a porous material can vary independently of $\mathbf{E}_s$, and is therefore included as an argument in the internal energy of the material. The derivatives $\dfrac{de_f}{d\overline{\rho}_f}$, $\dfrac{\partial e_s}{\partial \overline{\rho}_s}$, and $\dfrac{\partial e_s}{\partial \mathbf{E}_s}$ will be assumed to exist and be continuous, and the fields $\operatorname{grad} \dfrac{de_f}{d\overline{\rho}_f}$, $\operatorname{grad} \dfrac{\partial e_s}{\partial \overline{\rho}_s}$, and $\operatorname{grad} \dfrac{\partial e_s}{\partial \mathbf{E}_s}$ will be assumed to be continuous on $\overline{B}_t \times [t_1, t_2]$.

To determine the variation of the potential energy, (4.64) is expressed in terms of the comparison motion (4.10) and the comparison material density $(4.12)_1$. Upon taking the derivative of the result with respect to ε and setting $\varepsilon = 0$, the variation is

$$\delta U = \int_{B_t} \left[\rho_f \frac{de_f}{d\overline{\rho}_f} \delta \overline{\rho}_f + \rho_s \frac{\partial e_s}{\partial \overline{\rho}_s} \delta \overline{\rho}_s - \operatorname{div} \left(\rho_s \frac{\partial e_s}{\partial \mathbf{E}_s} \mathbf{F}^t \right) \cdot \delta \mathbf{x}_s \right] dV_t. \tag{4.65}$$

Hamilton's principle for a mixture of an elastic, ideal fluid and an elastic solid states [5]: *Among comparison motions (4.10) and comparison fields (4.12), the actual fields are such that*

$$\int_{t_1}^{t_2} [\delta(T - U) + \delta W + \delta C] dt = 0. \tag{4.66}$$

The virtual work δW, the constraint term δC, and the kinetic energy T are given by the same expressions (4.25), (4.26), and (4.27) as in the case of a mixture of compressible fluids. Since the only change in Equation (4.29) is the replacement of the expression for the variation of the potential energy by (4.65), it is easy to show that the resulting differential equations are

$$\rho_s \mathbf{a}_s = \rho_s \mathbf{b}_s + \mathbf{d}_s - \operatorname{grad} \pi_s + \lambda \operatorname{grad} \phi_s + \operatorname{div} \left(\rho_s \frac{\partial e_s}{\partial \mathbf{E}_s} \mathbf{F}^t \right),$$

$$\rho_f \mathbf{a}_f = \rho_f \mathbf{b}_f + \mathbf{d}_f - \operatorname{grad} \pi_f + \lambda \operatorname{grad} \phi_f,$$

$$\pi_s = \phi_s \overline{\rho}_s^2 \frac{\partial e_s}{\partial \overline{\rho}_s},$$

$$\pi_f = \phi_f \overline{\rho}_f^2 \frac{de_f}{d\overline{\rho}_f}, \tag{4.67}$$

$$\pi_s = \phi_s \lambda,$$

$$\pi_f = \phi_f \lambda$$

on $\overline{B}_t \times [t_1, t_2]$. Upon using the last two equations to eliminate π_s and π_f, the remaining equations can be written

$$\rho_s \mathbf{a}_s = \rho_s \mathbf{b}_s + \mathbf{d}_s - \phi_s \operatorname{grad} \lambda + \operatorname{div}\left(\rho_s \frac{\partial e_s}{\partial \mathbf{E}_s} \mathbf{F}^t\right),$$

$$\rho_f \mathbf{a}_f = \rho_f \mathbf{b}_f + \mathbf{d}_f - \phi_f \operatorname{grad} \lambda, \tag{4.68}$$

$$\lambda = \bar{\rho}_s^2 \frac{\partial e_s}{\partial \bar{\rho}_s} = \bar{\rho}_f^2 \frac{d e_f}{d \bar{\rho}_f}.$$

Let the material densities and the volume fractions be expressed as sums of their reference values and small perturbations,

$$\bar{\rho}_\xi = \bar{\rho}_{\xi R} + \tilde{\bar{\rho}}_\xi,$$

$$\phi_\xi = \phi_{\xi R} + \tilde{\phi}_\xi. \tag{4.69}$$

The linearized form of the equation of conservation of mass (4.7) is

$$\frac{\tilde{\phi}_\xi}{\phi_{\xi R}} + \frac{\tilde{\bar{\rho}}_\xi}{\bar{\rho}_{\xi R}} + \operatorname{tr} \mathbf{E}_\xi = 0, \tag{4.70}$$

and the linearized form of the volume fraction constraint (4.9) is

$$\sum_\xi \tilde{\phi}_\xi = 0. \tag{4.71}$$

The two volume fractions can be eliminated from the three equations (4.70) and (4.71) to obtain the single equation

$$\phi_{sR}\left(\frac{\tilde{\bar{\rho}}_s}{\bar{\rho}_{sR}} + \operatorname{tr} \mathbf{E}_s\right) + \phi_{fR}\left(\frac{\tilde{\bar{\rho}}_f}{\bar{\rho}_{fR}} + \operatorname{tr} \mathbf{E}_f\right) = 0. \tag{4.72}$$

Next, let the internal energies of the constituents be expressed as isotropic second-order expansions in their arguments.

$$\rho_s e_s = \tfrac{1}{2}\bar{c}(\operatorname{tr} \mathbf{E}_s)^2 + \bar{d}\,\mathbf{E}_s \cdot \mathbf{E}_s + \bar{f}\,\tilde{\bar{\rho}}_s \operatorname{tr} \mathbf{E}_s + \tfrac{1}{2}\bar{g}\,\tilde{\bar{\rho}}_s^2,$$

$$\rho_f e_f = \tfrac{1}{2}\bar{h}\,\tilde{\bar{\rho}}_f^2, \tag{4.73}$$

where $\bar{c}$, $\bar{d}$, $\bar{f}$, $\bar{g}$, and $\bar{h}$ are constitutive constants. Using these two representations, the linearized forms of Equations (4.68) can be written, with the external body forces neglected,

$$\rho_{sR}\ddot{\mathbf{u}}_s = \mathbf{d}_s - \phi_{sR}\,\text{grad}\,\lambda + (\overline{c} + 2\overline{d})\,\text{grad}\,\text{div}\,\mathbf{u}_s$$

$$-\overline{d}\,\text{curl}\,\text{curl}\,\mathbf{u}_s + \overline{f}\,\text{grad}\,\tilde{\rho}_s,$$

$$\rho_{fR}\ddot{\mathbf{u}}_f = \mathbf{d}_f - \phi_{fR}\,\text{grad}\,\lambda, \qquad\qquad (4.74)$$

$$\lambda = \frac{\overline{\rho}_{sR}}{\phi_{sR}}(\overline{f}\,\text{tr}\,\mathbf{E}_s + \overline{g}\,\tilde{\rho}_s),$$

$$\lambda = \frac{\overline{\rho}_{fR}}{\phi_{fR}}\overline{h}\,\tilde{\rho}_f.$$

When constitutive relations are specified for the drag terms $\mathbf{d}_\varsigma$, Equations (4.72) and (4.74) provide a linear system of equations with which to determine the fields $\mathbf{u}_s$, $\mathbf{u}_f$, $\tilde{\rho}_s$, $\tilde{\rho}_f$, and λ.

Equation (4.72) and the last two of Equations (4.74) can be solved for the variables $\tilde{\rho}_s$, $\tilde{\rho}_f$, and λ in terms of $\text{tr}\,\mathbf{E}_s$ and $\text{tr}\,\mathbf{E}_f$. When the resulting expressions are substituted into the first two of Equations (4.74), they can be written

$$\rho_{sR}\ddot{\mathbf{u}}_s = \mathbf{d}_s + (\overline{P} + 2\overline{N})\,\text{grad}\,\text{div}\,\mathbf{u}_s - \overline{N}\,\text{curl}\,\text{curl}\,\mathbf{u}_s + \overline{Q}\,\text{grad}\,\text{div}\,\mathbf{u}_f,$$

$$\rho_{fR}\ddot{\mathbf{u}}_f = \mathbf{d}_f + \overline{Q}\,\text{grad}\,\text{div}\,\mathbf{u}_s + \overline{R}\,\text{grad}\,\text{div}\,\mathbf{u}_f, \qquad (4.75)$$

where $\overline{P}$, $\overline{Q}$, $\overline{R}$, and $\overline{N}$ are constants. These two linear equations for the displacement fields $\mathbf{u}_s$ and $\mathbf{u}_f$ are the *Biot equations* [9].

The application of Hamilton's principle to a mixture of an elastic ideal fluid and an elastic solid described in this section has been extended to include microkinetic energy of the constituents by Bedford and Drumheller [5]. This approach has been used to develop a theory of a porous elastic material containing a bubbly liquid by Bedford and Stern [7], and it has been used to obtain a theory of a saturated porous medium with microstructure and nonlinear material behavior by Berryman and Thigpen [8].

4.4 A Theory of Mixtures with Microstructure

The theories discussed in Sections 4.2 and 4.3 are very special for the same reason that the theories of elastic ideal fluids and elastic solids described in Subsections 3.1.1 and 3.1.2 were special; internal energies were introduced which were assumed to depend only upon the states of deformation of the constituents. This restriction can be removed in the case of a mixture by using the same approach that was used in Subsection 3.1.3 for a single material. That is, internal forces can be expressed through virtual work terms rather than by internal energies. In this section, an illustration will be given of the use of Hamilton's principle to derive a quite general theory of mixtures with microkinetic energy in which the constituents are not constrained to be ideal or elastic [22]. The theory will include the results of the preceding two sections as special cases.

The Microkinetic Energy. Consider a homogeneous sphere of material of radius R and density $\bar{\rho}$. If the sphere expands homogeneously, it is easy to show that its kinetic energy relative to the center of the sphere is

$$\tfrac{2}{5}\pi\bar{\rho}R^3\dot{R}^2,$$

where the dot denotes the time derivative. Since the radius of the sphere and the density of the material in a homogeneous expansion are related by

$$\bar{\rho}\,\tfrac{4}{3}\pi R^3 = \text{constant},$$

the kinetic energy can be expressed in terms of the density as

$$\frac{\tfrac{2}{45}\pi(\bar{\rho}_R)^{\tfrac{5}{3}}R_R^5}{(\bar{\rho})^{\tfrac{8}{3}}}\dot{\bar{\rho}}^2,$$

where $\bar{\rho}_R$ and R_R are reference values.

Suppose that a constituent of a mixture consists of a distribution of such spheres. Multiplying this expression for the kinetic energy by the number of spheres per unit volume $\phi/(\tfrac{4}{3}\pi R^3)$, the kinetic energy per unit volume due to expansion or contraction of the spheres is

$$\tfrac{1}{2}\rho\left[\frac{(\bar{\rho}_R)^{\tfrac{2}{3}}R_R^2}{15(\bar{\rho})^{\tfrac{8}{3}}}\right]\dot{\bar{\rho}}^2,$$

which is of the same functional form as Equation (4.46). Therefore, two examples of microkinetic energy, the energy of the liquid surrounding a distribution of oscillating bubbles, discussed in Subsection 4.2.3, and the energy due to the homogeneous expansion and contraction of a distribution of particles, can be included in the present model if it is assumed that each constituent C_ξ has a microkinetic energy per unit volume of the form

$$\tfrac{1}{2}\rho_\xi I_\xi(\bar{\rho}_\gamma)\dot{\bar{\rho}}_\xi^2.$$

The coefficient $I_\xi(\bar{\rho}_\gamma)$ is a constitutive function which is assumed to depend upon the material density of each constituent. The second partial derivatives of these functions will be assumed to exist and to be continuous. Therefore, the total kinetic energy of the mixture contained in B_t will be expressed in the form

$$T = \sum_\xi \int_{B_t} \tfrac{1}{2}\rho_\xi(\mathbf{v}_\xi \cdot \mathbf{v}_\xi + I_\xi \dot{\bar{\rho}}_\xi^2)dV_t. \tag{4.76}$$

Hamilton's Principle. The virtual work done on the mixture contained in B_t by internal forces will be assumed to have the form

$$\delta W = -\sum_\xi \int_B \mathbf{S}_\xi \cdot \delta \mathbf{F}_\xi dV - \sum_\xi \int_{B_t} \frac{\phi_\xi}{\bar{\rho}_\xi} p_\xi \delta\bar{\rho}_\xi dV_t. \tag{4.77}$$

That is, it is assumed that work is done on C_ξ when its deformation gradient changes and when its material density changes. These two variables can change independently of one another if a constituent consists of, for example, a porous material or a distribution of particles. The form of (4.77) is motivated by the form of the virtual work done by internal forces in a single inelastic material, Equation (3.65), and by the independent variables which appear in the internal energy for a mixture of an ideal fluid and an elastic solid, Equation (4.64). The generalized forces, the tensor fields $\mathbf{S}_\xi$ and the scalar fields p_ξ, are constitutive variables. It will be assumed that $\mathbf{S}_\xi$, DIV $\mathbf{S}_\xi$, p_ξ, and GRAD p_ξ are continuous on $\overline{B} \times [t_1, t_2]$.

The virtual work done by external forces will be assumed to be of the form (4.25), and the comparison motions, comparison material density fields, and comparison volume fraction fields are subject to the constraint (4.26).

Hamilton's principle for the mixture states: *Among comparison motions* (4.10) *and comparison fields* (4.12), *the actual fields are such that*

$$\int_{t_1}^{t_2} (\delta T + \delta W + \delta C)dt = 0. \tag{4.78}$$

By substitution of the expressions (4.25), (4.26), (4.76), and (4.77), and through the use of steps that are familiar from previous sections, (4.78) can be written

$$\begin{aligned}
\sum_\xi \int_{t_1}^{t_2} \int_{B_t} \Big[\ & (-\rho_\xi \mathbf{a}_\xi + \operatorname{div} \mathbf{T}_\xi - \operatorname{grad} \pi_\xi \\
& + \lambda \operatorname{grad} \phi_\xi + \rho_\xi \mathbf{b}_\xi + \mathbf{d}_\xi) \cdot \delta\mathbf{x}_\xi \\
& + \left(-\rho_\xi \dot{\overline{I_\xi \dot{\bar{\rho}}_\xi}} + \sum_\gamma \tfrac{1}{2}\rho_\gamma \frac{\partial I_\gamma}{\partial \bar{\rho}_\xi} \dot{\bar{\rho}}_\gamma^2 - \phi_\xi \frac{p_\xi}{\bar{\rho}_\xi} + \frac{\pi_\xi}{\bar{\rho}_\xi} \right) \delta\bar{\rho}_\xi \\
& + \left(\frac{\pi_\xi}{\phi_\xi} - \lambda \right) \delta\phi_\xi \Big] dV_t dt = 0,
\end{aligned} \tag{4.79}$$

where

$$\mathbf{T}_\xi = \frac{1}{J_\xi}\mathbf{S}_\xi\mathbf{F}_\xi^t \qquad (4.80)$$

is the Cauchy stress of C_ξ. The resulting differential equations are

$$\left.\begin{aligned}
\rho_\xi\mathbf{a}_\xi &= \operatorname{div}\mathbf{T}_\xi - \operatorname{grad}\pi_\xi + \lambda\operatorname{grad}\phi_\xi \\[2pt]
&\quad + \rho_\xi\mathbf{b}_\xi + \mathbf{d}_\xi, \\[6pt]
\rho_\xi\overline{I_\xi\dot{\bar\rho}_\xi} - \sum_\gamma\tfrac{1}{2}\rho_\gamma\frac{\partial I_\gamma}{\partial\bar\rho_\xi}\dot{\bar\rho}_\gamma^2 &= \frac{\pi_\xi}{\bar\rho_\xi} - \phi_\xi\frac{p_\xi}{\bar\rho_\xi}, \\[6pt]
\pi_\xi &= \phi_\xi\lambda
\end{aligned}\right\} \quad \text{on } B_t\times[t_1,t_2]. \qquad (4.81)$$

Using the last equation to eliminate π_ξ, the equations reduce to

$$\rho_\xi\mathbf{a}_\xi = \operatorname{div}\mathbf{T}_\xi - \phi_\xi\operatorname{grad}\lambda + \rho_\xi\mathbf{b}_\xi + \mathbf{d}_\xi,$$

$$\bar\rho_\xi^2\overline{I_\xi\dot{\bar\rho}_\xi} - \sum_\gamma\tfrac{1}{2}\rho_\gamma\frac{\bar\rho_\xi}{\phi_\xi}\frac{\partial I_\gamma}{\partial\bar\rho_\xi}\dot{\bar\rho}_\gamma^2 = \lambda - p_\xi. \qquad (4.82)$$

To obtain a complete mechanical theory, constitutive relations must be postulated for the generalized forces $\mathbf{T}_\xi$, $\mathbf{d}_\xi$, and p_ξ and for the microkinetic energy coefficients I_ξ. Then (4.6), (4.8), (4.9), and (4.82) provide a system of equations to determine the fields ρ_ξ, $\bar\rho_\xi$, ϕ_ξ, λ, and $\mathbf{v}_\xi$.

Balance of Energy. A postulate of the equations of balance of energy for the mixture can be motivated by using the method described in Subsection 3.1.3. Consider an arbitrary volume B_t' contained within B_t (Figure 3.1). The part of C_ξ which is contained in B_t' at time t occupies a volume B_ξ' in the reference configuration. Recall the correspondence between the form of the virtual work term (3.65) and the mechanical working term (3.82) which appears in the global form of the equation of balance of energy in the case of an ordinary continuous medium. In the case of the mixture, the virtual work term is Equation (4.77). From the form of this term, it can be deduced that the mechanical working term for the part of C_ξ contained in B_t' is

$$\int_{B_\xi'}\mathbf{S}_\xi\cdot\dot{\mathbf{F}}_\xi\,dV_\xi + \int_{B_t'}\frac{\phi_\xi}{\bar\rho_\xi}p_\xi\dot{\bar\rho}_\xi\,dV_t.$$

Equating this expression to the rate of change of the internal energy of C_ξ within B_t' and introducing heat conduction terms analogous to (3.78) and (3.79), the balance of energy postulate for C_ξ is

$$\frac{d}{dt}\int_{B'_t}\rho_\xi e_\xi dV_t = \int_{B'_t}\mathbf{T}_\xi\cdot\mathbf{L}_\xi dV_t + \int_{B'_t}\frac{\phi_\xi}{\bar{\rho}_\xi}p_\xi\dot{\bar{\rho}}_\xi dV_t$$
$$-\int_{\partial B'_t}\mathbf{q}_\xi\cdot\mathbf{n}dS_t + \int_{B'_t}\rho_\xi s_\xi dV_t,$$

(4.83)

where Equation (3.82) has been used. The heat flux $\mathbf{q}_\xi$ is assumed to be C^1 and the heat supply s_ξ is assumed to be C^0 on $\overline{B}\times[t_1,t_2]$. Here the heat supply s_ξ is defined to be the rate at which heat is added to C_ξ both by external sources and by the other constituent of the mixture. This postulate implies the local form of the equation of balance of energy for C_ξ.

$$\rho_\xi\dot{e}_\xi = \mathbf{T}_\xi\cdot\mathbf{L}_\xi + \frac{\phi_\xi}{\bar{\rho}_\xi}p_\xi\dot{\bar{\rho}}_\xi - \operatorname{div}\mathbf{q}_\xi + \rho_\xi s_\xi.$$

(4.84)

Let the field $\theta_\xi(\mathbf{X}_\xi,t)$ denote the absolute temperature of C_ξ. Then if constitutive relations are postulated for $\mathbf{T}_\xi$, $\mathbf{d}_\xi$, p_ξ, I_ξ, e_ξ, $\mathbf{q}_\xi$, and s_ξ, Equations (4.6), (4.8), (4.9), (4.82), and (4.84) can be used to determine the fields ρ_ξ, $\bar{\rho}_\xi$, ϕ_ξ, λ, $\mathbf{v}_\xi$, and θ_ξ, yielding a *thermomechanical theory of mixtures with microkinetic energy.*

Independently, Nunziato, Passman, and Walsh ([59], [62], [64]) have developed a theory of mixtures with microstructure which shares many elements in common with this one. Their theory is based on the theory of granular solids due to Goodman and Cowin that is described in Subsection 3.2.1. They proceeded by adopting Equations (3.101) and (3.104) for each constituent of the mixture. They then introduced appropriate terms to account for the interchanges of momentum and energy between constituents.

5 Discontinuous fields

A *singular surface* is a surface across which some of the fields which characterize a continuous medium, or their derivatives, have jump discontinuities (see *e.g.* Truesdell and Toupin [71], pp. 491–529). The fields on either side of the surface are related by *jump conditions* which insure conservation of mass and balance of momentum and energy across the surface. These jump conditions are used, for example, in the study of boundary conditions and the propagation of wave fronts.

Hamilton's principle has been applied to an ideal fluid containing a singular surface by Taub [67]. However, his treatment was restricted to the case in which the velocity of the fluid is continuous across the surface. In the next two sections, the notation and definitions required to apply Hamilton's principle to a continuous medium containing a singular surface are discussed, and the case of an elastic ideal fluid is treated as an example.

5.1 Singular Surfaces

Assume that the motion

$$\mathbf{x} = \chi(\mathbf{X}, t) \tag{5.1}$$

is one-one and C^0 on $\overline{B} \times [t_1, t_2]$.[1] Let Σ denote a fixed, plane, open surface in $\mathcal{E}$, and define a function

$$\mathbf{z} = \varsigma(\mathbf{W}, t) \tag{5.2}$$

which maps Σ onto a surface Σ_t which intersects $\overline{B}_t$ at time t (Figure 5.1). The vector $\mathbf{W}$ denotes the position vector of a point of Σ (a *surface point*). The vector $\mathbf{z}$ is the position of the surface point $\mathbf{W}$ at time t. The mapping (5.2) will be assumed to be C^2 on $\Sigma \times [t_1, t_2]$. Let the intersection of Σ_t with $\overline{B}_t$ be denoted by S_t. The surface Σ does not represent a physical surface, but simply provides a means to describe the motion of the surface S_t. The surface S_t may represent a wave front or other surface of interest in the material at time t.

---

[1]See the discussion of the motion in Section 2.2.

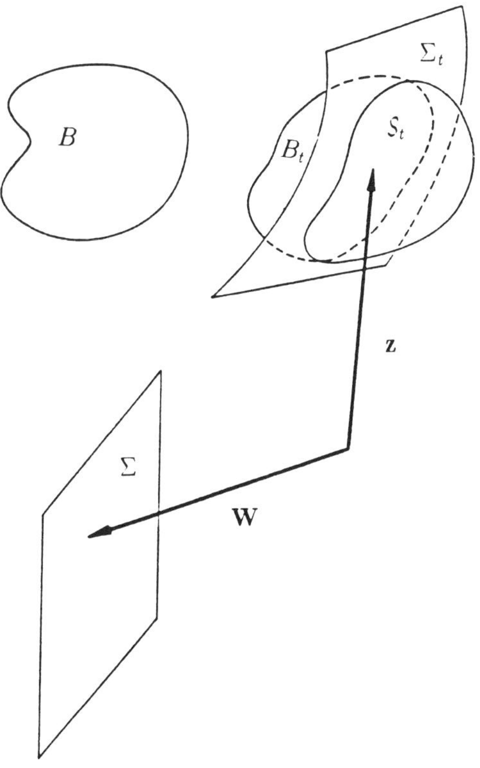

Figure 5.1: The surfaces Σ, Σ_t, and $\mathcal{S}_t$.

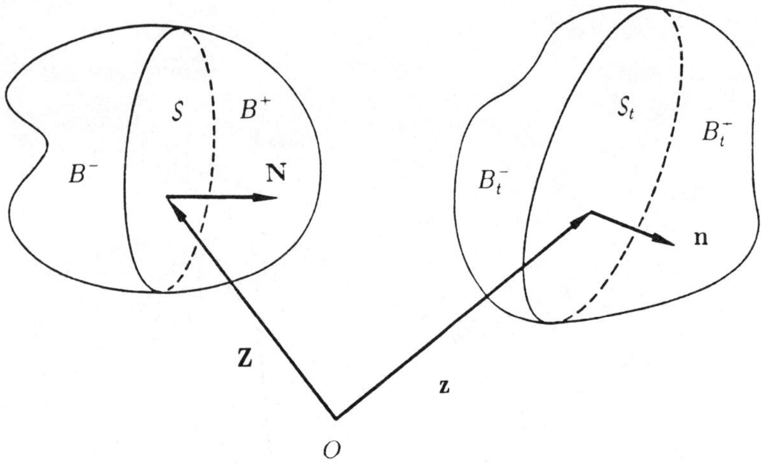

Figure 5.2: The surface S_t and the image surface S.

The surface S_t divides B_t into two parts which will be called B_t^+ and B_t^-. Let the field $\mathbf{n}(\mathbf{x}, t)$ defined on S_t be the unit vector normal to S_t which points into B_t^+ (Figure 5.2).

Since the motion of the material is assumed to be one-one and continuous on $\overline{B} \times [t_1, t_2]$, a unique material point is located at a given point $\mathbf{z}$ of S_t at time t. The position of this material point in the reference configuration, denoted by $\mathbf{Z}$, is given by the inverse motion,

$$\mathbf{Z} = \boldsymbol{\chi}^{-1}(\mathbf{z}, t). \tag{5.3}$$

This function maps the surface S_t onto a surface S in the reference configuration. The surface S is called the *image surface*; it is the image in the reference configuration of the surface S_t at time t. It divides the reference configuration into two parts B^+ and B^-. The function $\mathbf{N}(\mathbf{X}, t)$ defined on S will denote the unit vector normal to S which points into B^+.

Let ∂B^+ be the outer surface of B^+, and let $\overline{B}^+$ denote the closure of B^+; that is, B^+ together with its surface $\partial B^+ + S$. The notations ∂B^- and $\overline{B}^-$ are defined correspondingly. The motion (5.1) will be assumed to be C^2 on $\overline{B}^+ \times [t_1, t_2]$ and $\overline{B}^- \times [t_1, t_2]$. Thus *the motion is assumed to be C^2 on each part of B, but is merely assumed to be continuous across the surface S.*

90

Consider a field $\mathbf{f}(\mathbf{X}, t)$, and define $\mathbf{f}^+$ by

$$\mathbf{f}^+ = \lim_{\mathbf{X}\to\mathbf{Z}} \mathbf{f}(\mathbf{X}, t), \tag{5.4}$$

where the limit is taken as $\mathbf{X}$ approaches $\mathbf{Z}$ along a smooth path within B^+. The notation $\mathbf{f}^-$ is defined correspondingly. With some exceptions that will be obvious from their contexts, the superscripts $^+$ and $^-$ will refer to these limits.

The *jump* of $\mathbf{f}(\mathbf{X}, t)$ across S is defined by

$$[\![\mathbf{f}]\!] = \mathbf{f}^+ - \mathbf{f}^-. \tag{5.5}$$

The velocity of the surface point $\mathbf{W}$ is

$$\dot{\mathbf{z}} = \frac{\partial}{\partial t}\varsigma(\mathbf{W}, t). \tag{5.6}$$

The normal component $\dot{\mathbf{z}} \cdot \mathbf{n}$ is the speed of the surface S_t. It is called the *speed of displacement* ([71], p. 499). From (5.3), the velocity of the surface point $\mathbf{W}$ in the reference configuration is

$$\dot{\mathbf{Z}} = \frac{\partial}{\partial t}\chi^{-1}(\varsigma(\mathbf{W}, t), t)$$

$$= \left(\frac{\partial\chi^{-1}}{\partial\mathbf{X}}\right)^+ \frac{\partial\varsigma}{\partial t} + \left(\frac{\partial\chi^{-1}}{\partial t}\right)^+ \tag{5.7}$$

$$= \left(\frac{\partial\chi^{-1}}{\partial\mathbf{X}}\right)^- \frac{\partial\varsigma}{\partial t} + \left(\frac{\partial\chi^{-1}}{\partial t}\right)^-.$$

In this expression, the chain rule can be written in terms of the limits of the derivatives at S as a consequence of *Hadamard's lemma* ([71] pp. 492–505). Note that

$$d\mathbf{X} = \frac{\partial\chi^{-1}}{\partial\mathbf{x}}d\mathbf{x} + \frac{\partial\chi^{-1}}{\partial t}dt, \tag{5.8}$$

so the partial derivative of the inverse motion with respect to time can be written

$$\frac{\partial\chi^{-1}}{\partial t} = -\frac{\partial\chi^{-1}}{\partial\mathbf{x}}\left[\frac{d\mathbf{x}}{dt}\right]_{\mathbf{X}} = -\mathbf{F}^{-1}\mathbf{v}. \tag{5.9}$$

Upon substituting this result into Equation (5.7), a relation is obtained between the velocity of the material point $\mathbf{W}$ and the velocity of its image in the reference configuration.

$$\dot{\mathbf{Z}} = \left(\mathbf{F}^{-1}\right)^+ \left(\dot{\mathbf{z}} - \mathbf{v}^+\right)$$

$$= \left(\mathbf{F}^{-1}\right)^- \left(\dot{\mathbf{z}} - \mathbf{v}^-\right). \tag{5.10}$$

The normal component $\dot{\mathbf{Z}} \cdot \mathbf{N}$ is the speed of the surface S. It is called the *speed of propagation* ([71], p. 508). Equation (5.10) yields the expression

$$[\![\mathbf{F}^{-1}(\dot{\mathbf{z}} - \mathbf{v})]\!] = 0. \tag{5.11}$$

This equation results from the assumed continuity of the motion (5.1) across S. Taking the inner product of it with $\mathbf{N}$ and using (2.54) and (2.59) yields the well known result

$$[\![\rho(\dot{\mathbf{z}} - \mathbf{v}) \cdot \mathbf{n}]\!] = 0. \tag{5.12}$$

This equation is called the *Stokes-Christoffel condition* ([71], p. 522). It is the jump condition which insures conservation of mass of the material across S_t.

The comparison motion will be defined by[2]

$$\begin{aligned}
\mathbf{x}^* &= \boldsymbol{\chi}(\mathbf{X}, t) + \varepsilon\boldsymbol{\eta}(\mathbf{X}, t) \\
&= \mathbf{K}(\mathbf{X}, t, \varepsilon).
\end{aligned} \tag{5.13}$$

The vector field $\boldsymbol{\eta}(\mathbf{X}, t)$ is arbitrary subject to the conditions that it be C^2 on $\overline{B}^+ \times [t_1, t_2]$ and on $\overline{B}^- \times [t_1, t_2]$, that $\boldsymbol{\eta}(\mathbf{X}, t_1) = \mathbf{o}$ and $\boldsymbol{\eta}(\mathbf{X}, t_2) = \mathbf{o}$, and that (5.13) satisfy prescribed boundary conditions on ∂B.[3] The field $\boldsymbol{\eta}(\mathbf{X}, t)$ is *not* assumed to be continuous across S.

The comparison density field will be defined by

$$\rho^* = \rho(\mathbf{X}, t) + \varepsilon r(\mathbf{X}, t), \tag{5.14}$$

where $r(\mathbf{X}, t)$ is an arbitrary scalar field subject to the conditions that it be C^1 on $\overline{B}^+ \times [t_1, t_2]$ and on $\overline{B}^- \times [t_1, t_2]$ and that $r(\mathbf{X}, t_1) = 0$ and $r(\mathbf{X}, t_2) = 0$. The field $r(\mathbf{X}, t)$ is also not assumed to be continuous across S.

In analogy with (5.13), a comparison motion of the surface S_t is defined by [67]

$$\mathbf{z}^* = \boldsymbol{\varsigma}(\mathbf{W}, t) + \varepsilon\boldsymbol{\mu}(\mathbf{W}, t), \tag{5.15}$$

where $\boldsymbol{\mu}(\mathbf{W}, t)$ is an arbitrary C^2 function on $\Sigma \times [t_1, t_2]$ such that $\boldsymbol{\mu}(\mathbf{W}, t_1) = \mathbf{o}$ and $\boldsymbol{\mu}(\mathbf{W}, t_2) = \mathbf{o}$.

---

[2]See the discussion of the comparison motion in Section 2.3.

[3]Since the example to be presented is an ideal fluid, and there is no concern with boundary conditions on ∂B, it will be assumed henceforth that $\boldsymbol{\eta} \cdot \mathbf{n} = 0$ on ∂B.

92

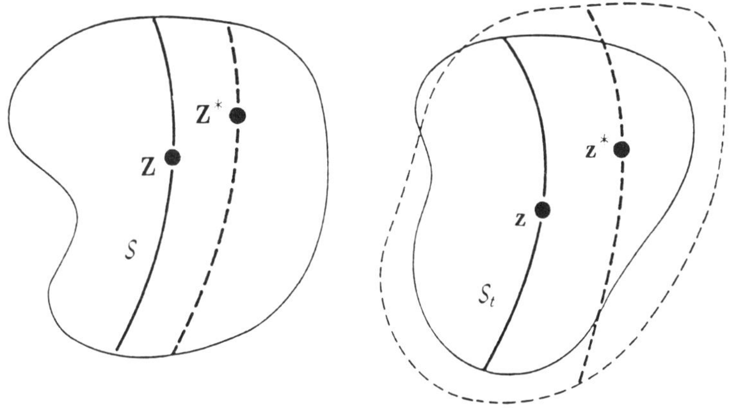

Figure 5.3: The points $\mathbf{z}^*$ and $\mathbf{Z}^*$.

As a result of the comparison motions (5.13) and (5.15), the position in the reference configuration of the material point which is located at $\mathbf{z}^*$ at time t is (Figure 5.3)

$$\mathbf{Z}^* = \mathbf{K}^{-1}(\mathbf{z}^*, t, \varepsilon). \tag{5.16}$$

Expanding this equation with respect to ε,

$$\begin{aligned}
\mathbf{Z}^* &= \mathbf{Z} + \left[\left(\frac{\partial \mathbf{K}^{-1}}{\partial \mathbf{z}^*} \right)^+ \frac{\partial \mathbf{z}^*}{\partial \varepsilon} + \left(\frac{\partial \mathbf{K}^{-1}}{\partial \varepsilon} \right)^+ \right]_{\varepsilon\,=\,0} \varepsilon + O\left(\varepsilon^2\right) \\
&= \mathbf{Z} + \left(\mathbf{F}^{-1}\right)^+ (\boldsymbol{\mu} - \boldsymbol{\eta}^+)\varepsilon + O\left(\varepsilon^2\right),
\end{aligned} \tag{5.17}$$

where the result (4.18) has been used. This equation also holds when the $^+$ superscripts are replaced by $^-$. Introducing the notation (2.74), Equation (5.17) yields the result

$$\begin{aligned}
\delta\mathbf{Z} &= \left(\mathbf{F}^{-1}\right)^+ \left(\delta\mathbf{z} - \delta\mathbf{x}^+\right) \\
&= \left(\mathbf{F}^{-1}\right)^- \left(\delta\mathbf{z} - \delta\mathbf{x}^-\right),
\end{aligned} \tag{5.18}$$

where $\delta\mathbf{z} = \boldsymbol{\mu}$. Therefore,

$$\left[\!\left[\mathbf{F}^{-1}(\delta\mathbf{z} - \delta\mathbf{x}) \right]\!\right] = \mathbf{o}. \tag{5.19}$$

This equation is a constraint imposed on the comparison motions (5.13) and (5.15) by the continuity of the motion at the surface S. Compare Equations (5.18) and

93

(5.19) to Equations (5.10) and (5.11).

Let $\mathbf{f}(\mathbf{X},t)$ be a field that is continuous on $\overline{B}^+ \times [t_1,t_2]$ and $\overline{B}^- \times [t_1,t_2]$, and let $\mathbf{f}^*(\mathbf{X},t,\varepsilon)$ be its associated comparison field. Consider the integral

$$I = \int_{B_t^\pm} \rho \mathbf{f}\, dV_t = \int_{B\pm} \rho_R \mathbf{f}\, dV, \tag{5.20}$$

where the notation $B_t^\pm$ means the sum of the integrals over B_t^+ and B_t^-. When it is expressed in terms of the comparison field $\mathbf{f}^*(\mathbf{X},t,\varepsilon)$ and the comparison motions (5.13) and (5.15), the value of this integral is (see Figure 5.3)

$$I^*(\varepsilon) = \int_{B\pm} \rho_R \mathbf{f}^*\, dV - \int_S [\![\rho_R \mathbf{f}[(\mathbf{Z}^* - \mathbf{Z}) \cdot \mathbf{N}]]\!]\, dS + O(\varepsilon^2). \tag{5.21}$$

The second integral in this expression is due to the displacement of the surface S. Taking the derivative with respect to ε and equating ε to zero, the variation is

$$\delta I = \int_{B\pm} \rho_R \delta \mathbf{f}\, dV - \int_S [\![\rho_R \mathbf{f}(\delta \mathbf{Z} \cdot \mathbf{N})]\!]\, dS. \tag{5.22}$$

By using the results (2.59) and (5.18), this variation can be expressed in terms of integrals over B_t and S_t.

$$\delta I = \int_{B_t^\pm} \rho \delta \mathbf{f}\, dV_t - \int_{S_t} [\![\rho \mathbf{f}(\delta \mathbf{z} - \delta \mathbf{x}) \cdot \mathbf{n}]\!]\, dS_t. \tag{5.23}$$

As an example of the application of these results, the kinetic energy is[4]

$$T = \int_{B_t^\pm} \tfrac{1}{2}\rho \mathbf{v} \cdot \mathbf{v}\, dV_t = \int_{B\pm} \tfrac{1}{2}\rho_R \mathbf{v} \cdot \mathbf{v}\, dV. \tag{5.24}$$

Consider the integral of the kinetic energy from t_1 to t_2,

$$I = \int_{t_1}^{t_2} T\, dt = \int_{t_1}^{t_2} \int_{B\pm} \tfrac{1}{2}\rho_R \mathbf{v} \cdot \mathbf{v}\, dV\, dt. \tag{5.25}$$

From Equation (5.22), the variation of this integral is

$$\delta I = \int_{t_1}^{t_2} \bigg[\int_{B\pm} \rho_R \mathbf{v} \cdot \dot{\boldsymbol{\eta}}\, dV \\ - \int_S [\![\tfrac{1}{2}\rho_R(\mathbf{v} \cdot \mathbf{v})(\delta \mathbf{Z} \cdot \mathbf{N})]\!]\, dS \bigg]\, dt. \tag{5.26}$$

In integrating the first term by parts, the motion of the surface S must be accounted for. This can be done by evaluating the derivative

$$\frac{d}{dt} \int_{B\pm} \rho_R \mathbf{v} \cdot \boldsymbol{\eta}\, dV = \int_{B\pm} \rho_R \mathbf{a} \cdot \boldsymbol{\eta}\, dV + \int_{B\pm} \rho_R \mathbf{v} \cdot \dot{\boldsymbol{\eta}}\, dV \\ - \int_S [\![\rho_R(\mathbf{v} \cdot \boldsymbol{\eta})(\dot{\mathbf{Z}} \cdot \mathbf{N})]\!]\, dS. \tag{5.27}$$

---

[4]See the treatment of this example on page 28.

Upon integrating this equation with respect to time from t_1 to t_2 and using the assumption that $\boldsymbol{\eta}$ vanishes at t_1 and t_2, one obtains the required integration by parts

$$\int_{t_1}^{t_2}\int_{B\pm}\rho_R\mathbf{v}\cdot\dot{\boldsymbol{\eta}}\,dV\,dt = \int_{t_1}^{t_2}\left[-\int_{B\pm}\rho_R\mathbf{a}\cdot\boldsymbol{\eta}\,dV + \int_S[\![\rho_R(\mathbf{v}\cdot\boldsymbol{\eta})(\dot{\mathbf{Z}}\cdot\mathbf{N})]\!]\,dS\right]dt. \quad (5.28)$$

Using this result, the variation of the kinetic energy is

$$\begin{aligned}
\delta T =\ & -\int_{B\pm}\rho_R\mathbf{a}\cdot\delta\mathbf{x}\,dV + \int_S[\![(\rho_R\mathbf{v}\otimes\dot{\mathbf{Z}})\mathbf{N}\cdot\delta\mathbf{x}]\!]\,dS \\
& -\int_S[\![\tfrac{1}{2}\rho_R(\mathbf{v}\cdot\mathbf{v})\mathbf{N}\cdot\delta\mathbf{Z}]\!]\,dS \\
=\ & -\int_{B_t^\pm}\rho\mathbf{a}\cdot\delta\mathbf{x}\,dV_t + \int_{S_t}[\![[\rho\mathbf{v}\otimes(\dot{\mathbf{z}}-\mathbf{v})]\mathbf{n}\cdot\delta\mathbf{x}]\!]\,dS_t \\
& -\int_{S_t}[\![\tfrac{1}{2}\rho(\mathbf{v}\cdot\mathbf{v})\mathbf{n}\cdot(\delta\mathbf{z}-\delta\mathbf{x})]\!]\,dS_t,
\end{aligned} \qquad (5.29)$$

where the relations (2.59), (5.10), and (5.18) have been used.

As a second example, consider the constraint term associated with the equation of conservation of mass[5]

$$C = \int_{B\pm}\pi\left(J - \frac{\rho_R}{\rho}\right)dV = \int_{B_t^\pm}\pi\left(1 - \frac{\rho_R}{\rho J}\right)dV_t. \qquad (5.30)$$

From Equation (5.22), the variation is

$$\delta C = \int_{B\pm}\pi J\left(\operatorname{div}\boldsymbol{\eta} + \frac{r}{\rho}\right)dV = \int_{B_t^\pm}\pi\left(\operatorname{div}\boldsymbol{\eta} + \frac{r}{\rho}\right)dV_t. \qquad (5.31)$$

The presence of the singular surface must be taken into account in applying the divergence theorem to the first term. When this is done, the variation can be written

$$\delta C = \int_{B_t^\pm}\left(-\operatorname{grad}\pi\cdot\delta\mathbf{x} + \frac{\pi}{\rho}\delta\rho\right)dV_t - \int_{S_t}[\![\pi\mathbf{n}\cdot\delta\mathbf{x}]\!]\,dS_t. \qquad (5.32)$$

The results that have been discussed in this section are quite general, and could be applied to any of the examples in Chapter 3. In the next section, their use will be illustrated using the specific case of an elastic ideal fluid.

---

[5]See the treatment of this example on page 29.

5.2   An Ideal Fluid Containing a Singular Surface

Consider a finite amount of an elastic ideal fluid that occupies a bounded regular region B at time t_1.[6] Let it be assumed that during the time interval $[t_1, t_2]$, the volume B_t is divided into two parts B_t^+ and B_t^- by a singular surface S_t. Hamilton's principle states: *Among comparison motions (5.13) and comparison density fields (5.14), the actual fields are such that*

$$\int_{t_1}^{t_2} [\delta(T - U) + \delta W + \delta C]\,dt = 0, \tag{5.33}$$

where

$$T = \int_{B_t^\pm} \tfrac{1}{2}\rho \mathbf{v} \cdot \mathbf{v}\,dV_t,$$

$$U = \int_{B_t^\pm} \rho e(\rho)\,dV_t,$$

$$\delta W = \int_{B_t^\pm} \rho \mathbf{b} \cdot \delta \mathbf{x}\,dV_t, \tag{5.34}$$

$$\delta C = \int_{B_t^\pm} \left(-\operatorname{grad} \pi \cdot \delta \mathbf{x} + \frac{\pi}{\rho}\delta\rho \right) dV_t - \int_{S_t} [\![\pi \mathbf{n} \cdot \delta \mathbf{x}]\!]\,dS_t$$
$$+ \int_{S_t} \boldsymbol{\nu} \cdot [\![\mathbf{F}^{-1}(\delta \mathbf{z} - \delta \mathbf{x})]\!]\,dS_t.$$

The constraint term δC contains both the constraint arising from the equation of conservation of mass, Equation (5.32), and the constraint imposed by the continuity of the motion at S, Equation (5.19). The vector field $\boldsymbol{\nu}(\mathbf{z}, t)$ is a Lagrange multiplier that is assumed to be continuous on $S_t \times [t_1, t_2]$.

By using the expressions (5.23) and (5.29), Equation (5.33) can be written

$$\int_{t_1}^{t_2} \Bigg\{ -\int_{B_t^\pm} \rho \mathbf{a} \cdot \delta \mathbf{x}\,dV_t + \int_{S_t} [\![[\rho \mathbf{v} \otimes (\dot{\mathbf{z}} - \mathbf{v})]\mathbf{n} \cdot \delta \mathbf{x}]\!]\,dS_t$$
$$-\int_{S_t} [\![\tfrac{1}{2}\rho(\mathbf{v} \cdot \mathbf{v})\mathbf{n} \cdot (\delta \mathbf{z} - \delta \mathbf{x})]\!]\,dS_t$$
$$-\int_{B_t^\pm} \rho \frac{de}{d\rho}\delta\rho\,dV_t + \int_{S_t} [\![\rho e(\delta \mathbf{z} - \delta \mathbf{x}) \cdot \mathbf{n}]\!]\,dS_t \tag{5.35}$$
$$+\int_{B_t^\pm} \rho \mathbf{b} \cdot \delta \mathbf{x}\,dV_t + \int_{B_t^\pm} \left(-\operatorname{grad} \pi \cdot \delta \mathbf{x} + \frac{\pi}{\rho}\delta\rho \right) dV_t$$
$$-\int_{S_t} [\![\pi \mathbf{n} \cdot \delta \mathbf{x}]\!]\,dS_t + \int_{S_t} \boldsymbol{\nu} \cdot [\![\mathbf{F}^{-1}(\delta \mathbf{z} - \delta \mathbf{x})]\!]\,dS_t \Bigg\}\,dt = 0.$$

---

[6]See the discussion of ideal fluids in Subsection 3.1.1.

If it is assumed that $\delta\mathbf{z} = \mathbf{o}$ and that the variations $\delta\mathbf{x}^{\pm}$ vanish on S_t, this equation reduces to the case considered in Subsection 3.1.1 and yields the equation of balance of linear momentum (3.16) on $\overline{B}^+ \times [t_1, t_2]$ and on $\overline{B}^- \times [t_1, t_2]$. As a consequence, only the terms involving integrals over S_t remain in Equation (5.35). Next, let $\delta\mathbf{x}^{\pm} = \mathbf{o}$ while $\delta\mathbf{z}$ is permitted to be arbitrary on S_t. This results in the jump condition

$$\llbracket -\tfrac{1}{2}\rho(\mathbf{v} \cdot \mathbf{v})\mathbf{n} + \rho e\mathbf{n} + \mathbf{F}^{-t}\boldsymbol{\nu}\rrbracket = \mathbf{o} \quad \text{on } S_t \times [t_1, t_2]. \tag{5.36}$$

Finally, permitting the variations $\delta\mathbf{x}^+$ and $\delta\mathbf{x}^-$ to be arbitrary yields the two equations

$$\{[\rho\mathbf{v} \otimes (\dot{\mathbf{z}} - \mathbf{v})]\mathbf{n} + \tfrac{1}{2}\rho(\mathbf{v} \cdot \mathbf{v})\mathbf{n}$$
$$-\rho e\mathbf{n} - \pi\mathbf{n} - \mathbf{F}^{-t}\boldsymbol{\nu}\}^{\pm} = \mathbf{o}. \tag{5.37}$$

Subtracting the $^-$ equation from the $^+$ equation and adding the result to Equation (5.36) leads to the jump condition

$$\llbracket [\rho\mathbf{v} \otimes (\dot{\mathbf{z}} - \mathbf{v})]\mathbf{n} - \pi\mathbf{n}\rrbracket = \mathbf{o} \quad \text{on } S_t \times [t_1, t_2]. \tag{5.38}$$

This is the *momentum jump condition*. It insures conservation of linear momentum of the material across S_t (see *e.g.* [27], pp. 104–106).

Eliminating the Lagrange multiplier $\boldsymbol{\nu}$ from the two equations (5.37) results in the jump condition

$$\llbracket \mathbf{F}^t\{[\rho\mathbf{v} \otimes (\dot{\mathbf{z}} - \mathbf{v})]\mathbf{n} + \tfrac{1}{2}\rho(\mathbf{v} \cdot \mathbf{v})\mathbf{n}$$
$$-\rho e\mathbf{n} - \pi\mathbf{n}\}\rrbracket = \mathbf{o} \quad \text{on } S_t \times [t_1, t_2]. \tag{5.39}$$

Taking the inner product of this equation with $\dot{\mathbf{Z}}$, taking the inner product of (5.38) with $\dot{\mathbf{z}}$, and summing the results leads with some rearrangement to the usual form of the *energy jump condition*

$$\llbracket \rho(e + \tfrac{1}{2}\mathbf{v} \cdot \mathbf{v})[(\dot{\mathbf{z}} - \mathbf{v}) \cdot \mathbf{n}] - \pi\mathbf{v} \cdot \mathbf{n}\rrbracket = 0 \quad \text{on } S_t \times [t_1, t_2]. \tag{5.40}$$

This equation insures conservation of energy of the material across S_t (see *e.g.* [27], pp. 121–123). This derivation of the energy jump condition does not include terms associated with heat conduction.

Although (5.39) is more complicated than the usual form of the energy jump condition, note that in the case of one dimensional motion it reduces to the very simple form

$$\llbracket e + \tfrac{1}{2}v^2 - \dot{z}v + \pi/\rho\rrbracket = 0. \tag{5.41}$$

Thus Hamilton's principle yields both the linear momentum and the energy jump conditions for the fluid. This procedure has been extended to mixtures of fluids and elastic materials by Batra [3]. It could potentially be extended to other generalized theories of continuous media.

Acknowledgements

This monograph is a result of my fifteen year professional collaboration with my friend Douglas S. Drumheller of Sandia National Laboratories. Our work on mixtures led us to become interested in Hamilton's principle, and in addition to his original contributions to the research, Doug helped me to understand many of the subtleties of variational methods in continuum mechanics. The last two chapters are based in large part on our results. He is not a coauthor only because we did not have an opportunity to work together while it was being written. Our collaboration took place during periods when I was a temporary staff member and a consultant at Sandia National Laboratories. I am grateful for the consistently courteous and generous treatment I have received from that organization, and particularly want to thank James R. Asay, Walter Herrmann and Darrell E. Munson. I appreciate the support of my work there by the United States Department of Energy. Part of the work on which Chapter 5 is based was done in collaboration with my former graduate student Gautam Batra, and the monograph also benefitted from his ideas on notation and the organization of the subject. Part of the manuscript was written while I was a member of the technical staff of Applied Research Laboratories, The University of Texas at Austin. I thank the Office of Naval Research for their support of my work there. Gautam Batra, Ray M. Bowen, Douglas S. Drumheller, Morton E. Gurtin, Stephen L. Passman, Morris Stern, and Timothy G. Trucano read an early version of the manuscript and gave me many helpful suggestions. I also want to thank Jean G. Grissom and David B. Kaplan of Pitman Publishing Inc. for their help in the editorial process. The final manuscript was prepared at Sandia National Laboratories using the typesetting language LaTeX on a VAX 11/780 computer with an Imagen 24/300 printer.

References

[1] Akhiezer, N. I., *The Calculus of Variations*, Blaisdell Publishing Company, New York, 1962.

[2] Bateman, H., "Hamilton's work in dynamics and its influence on modern thought," *A Collection of Papers in Memory of Sir William Rowan Hamilton*, pp. 51–64, Scripta Mathematica, Yeshiva College, New York, 1945.

[3] Batra, G., *On the Propagation of One-Dimensional Waves of Discontinuity in Mixtures*, Ph. D. Dissertation, The University of Texas, Austin, Texas, 1984.

[4] Bedford, A., and Drumheller, D. S., "A variational theory of immiscible mixtures," *Archive for Rational Mechanics and Analysis* **68**, 37–51, 1978.

[5] Bedford, A., and Drumheller, D. S., "A variational theory of porous media," *International Journal of Solids and Structures* **15**, 967–980, 1979.

[6] Bedford, A., and Drumheller, D. S., "Theories of immiscible and structured mixtures," *International Journal of Engineering Science* **21**, 863–960, 1983.

[7] Bedford, A., and Stern, M., "A model for wave propagation in gassy sediments," *Journal of the Acoustical Society of America* **73**, 409–417, 1983.

[8] Berryman, J. G., and Thigpen, L., "Nonlinear and semilinear dynamic poroelasticity with microstructure," *Journal of the Mechanics and Physics of Solids* **33**, 97–116, 1985.

[9] Biot, M. A., "Theory of propagation of elastic waves in a fluid-saturated porous solid—I. Low frequency range," *Journal of the Acoustical Society of America* **28**, 168–178, 1956.

[10] Bliss, G. A., *Lectures on the Calculus of Variations*, University of Chicago Press, Chicago, Illinois, 1961.

[11] Bolza, O., *Lectures on the Calculus of Variations*, Chelsea Publishing Company, New York, 1973.

[12] Bowen, R. M., "Theory of Mixtures," *Continuum Physics Vol. III, Mixtures and EM Field Theories*, pp. 1–127, edited by A. C. Eringen, Academic Press, New York, 1976.

[13] Bowen, R. M., and Wang, C. -C., *Introduction to Vectors and Tensors Part A: Linear and Multilinear Algebra*, Plenum Press, New York, 1976.

[14] Bowen, R. M., and Wang, C. -C., *Introduction to Vectors and Tensors Part B: Vector and Tensor Analysis*, Plenum Press, New York, 1976.

[15] Courant, R., and Hilbert, D., *Methods of Mathematical Physics, Vol. I*, Interscience, New York, 1953.

[16] Cowin, S. C., "A theory for the flow of granular materials," *Powder Technology* **9**, 61–69, 1974.

[17] Cowin, S. C., and Goodman, M. A., "A variational principle for granular materials," *ZAMM* **56**, 281–286, 1976.

[18] Cowin, S. C., and Nunziato, J. W., "Waves of dilatancy in a granular material with incompressible grains," *International Journal of Engineering Science* **19**, 993–1008, 1981.

[19] Craine, R. E., "Oscillations of a plate in a binary mixture of incompressible Newtonian fluids," *International Journal of Engineering Science* **9**, 1177–1192, 1971.

[20] Drew, D. A., "Two-phase flows: Constitutive equations for lift and Brownian motion and some basic flows," *Archive for Rational Mechanics and Analysis* **62**, 149–163, 1976.

[21] Drumheller, D. S., and Bedford, A., "A theory of bubbly liquids," *Journal of the Acoustical Society of America* **66**, 197–208, 1979.

[22] Drumheller, D. S., and Bedford, A., "A thermomechanical theory for reacting immiscible mixtures," *Archive for Rational Mechanics and Analysis* **73**, 257–284, 1980.

[23] Drumheller, D. S., Kipp, M. E., and Bedford, A., "Transient wave propagation in bubbly liquids," *Journal of Fluid Mechanics* **119**, 347–365, 1982.

[24] Eckart, C., "Variational principles of hydrodynamics," *Physics of Fluids* **3**, 421–427, 1960.

[25] Ericksen, J. L., "Tensor fields," *Encyclopedia of Physics Vol. III/1*, edited by S. Flügge, pp. 794–858, Springer-Verlag, Berlin-Göttingen-Heidelberg, 1960.

[26] Ericksen, J. L., "Conservation laws for liquid crystals," *Transactions of the Society of Rheology* **5**, 23–34, 1961.

[27] Eringen, A. C., *Mechanics of Continua*, John Wiley, New York, 1967.

[28] Finlayson, B. A., *The Method of Weighted Residuals and Variational Principles*, Academic Press, New York, 1972.

[29] Funk, P., *Variationsrechnung und ihre Anwendung in Physik und Technik*, Springer-Verlag, Berlin-Göttingen-Heidelberg, 1962.

[30] Gelfand, I. M., and Fomin, S. V., *Calculus of Variations*, Prentice-Hall, Inglewood Cliffs, New Jersey, 1963.

[31] Goldstein, H., *Classical Mechanics*, Addison-Wesley, Reading, Massachusetts, 1959.

[32] Goodman, M. A., and Cowin, S. C., "A continuum theory for granular materials," *Archive for Rational Mechanics and Analysis* **44**, 249–266, 1972.

[33] Graves, R. P., *Life of Sir William Rowan Hamilton*, Longmans, Green & Company, London, 1882.

[34] Gurtin, M. E., "Variational principles for linear elastodynamics," *Archive for Rational Mechanics and Analysis* **16**, 34–50, 1964.

[35] Gurtin, M. E., "The linear theory of elasticity," *Encyclopedia of Physics Vol. VIa/2*, pp. 1–295, edited by C. Truesdell, Springer-Verlag, Berlin-Heidelberg-New York, 1972.

[36] Gurtin, M. E., *An Introduction to Continuum Mechanics*, Academic Press, New York, 1981.

[37] Halmos, P. R., *Finite-Dimensional Vector Spaces*, D. Van Nostrand, Princeton, New Jersey, 1958.

[38] Hamilton, W. R., "Theory of systems of rays," *Transactions of the Royal Irish Academy* **15**, 69–174, 1828.

[39] Hamilton, W. R., "On a general method in dynamics," *Philosophical Transactions of the Royal Society of London*, Part II, 247–308, 1834.

[40] Hamilton, W. R., "Second essay on a general method in dynamics," *Philosophical Transactions of the Royal Society of London*, Part I, 95–144, 1835.

[41] Hamilton, W. R., *The Mathematical Papers of Sir William Rowan Hamilton*, Vol. I, *Geometrical Optics*, edited by A. W. Conway and J. L. Synge, Cambridge University Press, London, 1931.

[42] Hamilton, W. R., *The Mathematical Papers of Sir William Rowan Hamilton*, Vol. II, *Dynamics*, edited by A. W. Conway and A. J. McConnell, Cambridge University Press, London. 1940.

[43] Hankins, T. L., *Sir William Rowan Hamilton*, Johns Hopkins University Press, Baltimore, Maryland, 1980.

[44] Herivel, J. W., "The derivation of the equations of motion of an ideal fluid by Hamilton's principle," *Proceedings of the Cambridge Philosophical Society* **51**, 344–349, 1955.

[45] Hill, C. D., "Two-dimensional analysis of the stability of particulate sedimentation," *Physics of Fluids* **23**, 667–668, 1980.

[46] Hill, C. D., and Bedford, A., "Stability of the equations for particulate sedimentation," *Physics of Fluids* **22**, 1252–1254, 1979.

[47] Hill, C. D., and Bedford, A., "A model for erythrocyte sedimentation," *Biorheology* **18**, 255–266, 1981.

[48] Hill, C. D., Bedford, A., and Drumheller, D. S., "An application of mixture theory to particulate sedimentation," *Journal of Applied Mechanics* **47**, 261–265, 1980.

[49] Kuznetsov, V. V., Nakoryakov, V. E., Pokusaev, B. G., and Shreiber, I. R., "Propagation of perturbations in a gas-liquid mixture," *Journal of Fluid Mechanics* **85**, 85–96, 1978.

[50] Lanczos, C., *The Variational Principles of Mechanics*, University of Toronto Press, Toronto, 1970.

[51] Leech, C. M., "Hamilton's principle applied to fluid mechanics," *Quarterly Journal of Mechanics and Applied Mathematics* **30**, 107–130, 1977.

[52] Leigh, D. C., *Nonlinear Continuum Mechanics*, McGraw-Hill, New York, 1968.

[53] Love, A. E. H., *A Treatise on the Mathematical Theory of Elasticity*, Dover, New York, 1944.

[54] Mindlin, R. D., "Micro-structure in linear elasticity," *Archive for Rational Mechanics and Analysis* **16**, 51–78, 1964.

[55] Nowinski, J. L., *Theory of Thermoelasticity with Applications*, Sijthoff and Noordhoff, The Netherlands, 1978.

[56] Nunziato, J. W., and Passman, S. L., "A multiphase mixture theory for fluid-saturated granular materials," *Mechanics of Structured Media A*, edited by A. P. S. Selvadurai, pp. 243–254, Elsevier, Amsterdam, 1981.

[57] Nunziato, J. W., Passman, S. L., and Thomas, J. P., "Gravitational flow of granular materials with incompressible grains," *Journal of Rheology* **24**, 395–420, 1980.

[58] Nunziato, J. W., and Walsh, E. K., "One-dimensional shock waves in uniformly distributed granular materials," *International Journal of Solids and Structures* **14**, 681–689, 1978.

[59] Nunziato, J. W., and Walsh, E. K., "On ideal multiphase mixtures with chemical reactions and diffusion," *Archive for Rational Mechanics and Analysis* **73**, 285–311, 1980.

[60] Oden, J. T., and Reddy, J. N., *Variational Methods in Theoretical Mechanics*, Springer-Verlag, Berlin-Heidelberg-New York, 1976.

[61] Pars, L. A., *An Introduction to the Calculus of Variations*, Heinemann, London, 1962.

[62] Passman, S. L., "Mixtures of granular materials," *International Journal of Engineering Science* **15**, 117–129, 1977.

[63] Passman, S. L., "Shearing flows of granular materials," *Journal of the Engineering Mechanics Division of the American Society of Civil Engineers* **106**, 773–783, 1980.

[64] Passman, S. L., Nunziato, J. W., and Walsh, E. K., "A theory of multiphase mixtures," appendix to *Rational Thermodynamics* by C. Truesdell, pp. 286–325, Springer-Verlag, New York-Berlin-Heidelberg-Tokyo, 1984.

[65] Serrin, J., "Mathematical principles of classical fluid mechanics," *Encyclopedia of Physics Vol. VIII/1*, edited by S. Flügge and C. Truesdell, pp. 125–263, Springer-Verlag, Berlin-Göttingen-Heidelberg, 1959.

[66] Silberman, E., "Sound velocity and attenuation in bubbly mixtures measured in standing wave tubes," *Journal of the Acoustical Society of America* **29**, 925–933, 1957.

[67] Taub, A. H., "On Hamilton's principle for perfect compressible fluids," *Nonlinear Problems in Mechanics of Continua*, edited by E. Reissner, W. Prager and J. J. Stoker, American Mathematical Society, New York, 1949.

[68] Torby, B. J., *Advanced Dynamics for Engineers*, Holt, Rinehart and Winston, New York, 1984.

[69] Truesdell, C., *Rational Thermodynamics*, Springer-Verlag, New York-Berlin-Heidelberg-Tokyo, 1984.

[70] Truesdell, C., and Noll, W., "The non-linear field theories of mechanics," *Encyclopedia of Physics Vol. III/3*, edited by S. Flügge, Springer-Verlag, Berlin-Heidelberg-New York, 1965.

[71] Truesdell, C., and Toupin, R. A., "The classical field theories," *Encyclopedia of Physics Vol. III/1*, edited by S. Flügge, pp. 226–793, Springer-Verlag, Berlin-Göttingen-Heidelberg, 1960.

[72] van Wijngaarden, L., "One-dimensional flow of liquids containing small gas bubbles," *Annual Reviews of Fluid Mechanics* **4**, 369–396, 1972.

[73] Washizu, K., *Variational Methods in Elasticity and Plasticity*, Pergamon Press, New York, 1982.

[74] Weinstock, R., *Calculus of Variations*, McGraw-Hill, New York, 1949.

[75] Whelan, J. A., Huang, C. -R., and Copley, A. L., "Concentration profiles in erythrocyte sedimentation in human whole blood," *Biorheology* **7**, 205–212, 1971.

[76] Whittaker, E. T., *A Treatise on the Analytical Dynamics of Particles and Rigid Bodies*, Dover, New York, 1944.